GUIDE TO LINEAR A

MACMILLAN MATHEMATICAL GUIDES

Consultant Editor: **David A. Towers**,
Senior Lecturer in Mathematics,
University of Lancaster

Linear Algebra D. Towers
Abstract Algebra C. Whitehead
Analysis F. Hart
Numerical Analysis J. Turner
Mathematical Modelling D. Edwards and M. Hanson
Mathematical Methods J. Gilbert
Mechanics P. Dyke and R. Whitworth

Guide to Linear Algebra

David A. Towers

Senior Lecturer in Mathematics
University of Lancaster

© David A. Towers 1988

All rights reserved. No reproduction, copy or transmission of this publication may be made without written permission.

No paragraph of this publication may be reproduced, copied or transmitted save with written permission or in accordance with the provisions of the Copyright, Designs and Patents Act 1988, or under the terms of any licence permitting limited copying issued by the Copyright Licensing Agency, 90 Tottenham Court Road, London W1P 9HE.

Any person who does any unauthorised act in relation to this publication may be liable to criminal prosecution and civil claims for damages.

First published 1988 by
THE MACMILLAN PRESS LTD
Houndmills, Basingstoke, Hampshire RG21 2XS
and London
Companies and representatives
throughout the world

ISBN 0-333-43627-X

A catalogue record for this book is available from the British Library.

12 11 10 9 8
03 02 01 00 99

Printed in Hong Kong

CONTENTS

Editor's foreword		viii
Preface		ix
1	**VECTORS**	**1**
	1.1 Free vectors and position vectors	1
	1.2 Bases and coordinates	6
	1.3 Scalar product	9
	1.4 Vector product	12
	1.5 Geometry of lines and planes in $\mathbb{R}^3$	14
	Solutions and hints for exercises	22
2	**MATRICES**	**28**
	2.1 Introduction	28
	2.2 Addition and scalar multiplication of matrices	30
	2.3 Matrix multiplication	33
	2.4 Properties of matrix multiplication	37
	2.5 The transpose of a matrix	40
	2.6 Invertible matrices	42
	Solutions and hints for exercises	45
3	**ROW REDUCTION**	**52**
	3.1 Systems of linear equations	52
	3.2 Equivalent systems	56
	3.3 Elementary row operations on matrices	60
	3.4 The reduced echelon form for a matrix	61
	3.5 Elementary matrices	65
	3.6 Finding the inverse of an invertible matrix	68
	Solutions and hints for exercises	70

4 DETERMINANTS — 74
- 4.1 The sign of a permutation — 74
- 4.2 The definition of a determinant — 76
- 4.3 Elementary properties of determinants — 79
- 4.4 Non-singular matrices — 85
- 4.5 Cofactors — 87
- 4.6 The adjugate of a matrix — 92
- 4.7 Systems of homogeneous linear equations — 94
- Solutions and hints for exercises — 96

5 VECTOR SPACES — 102
- 5.1 Introduction — 102
- 5.2 Subspaces and spanning sequences — 106
- 5.3 Linear independence and bases — 109
- 5.4 The dimension of a vector space — 112
- Solutions and hints for exercises — 117

6 LINEAR TRANSFORMATIONS — 124
- 6.1 Introduction — 124
- 6.2 Invertible linear transformations — 127
- 6.3 The matrix of a linear transformation — 130
- 6.4 Kernel and image of a linear transformation — 135
- 6.5 The rank of a matrix — 141
- 6.6 Systems of linear equations — 146
- Solutions and hints for exercises — 151

7 EIGENVECTORS — 160
- 7.1 Changing the domain basis — 160
- 7.2 Changing the codomain basis — 162
- 7.3 Changing the basis in both domain and codomain — 165
- 7.4 Eigenvalues and eigenvectors — 166
- 7.5 The characteristic equation of a square matrix — 169
- Solutions and hints for exercises — 172

8 ORTHOGONAL REDUCTION OF SYMMETRIC MATRICES — 177
- 8.1 Orthogonal vectors and matrices — 177
- 8.2 Euclidean transformations — 180
- 8.3 Orthogonal reduction of a real symmetric matrix — 184
- 8.4 Classification of conics — 192

8.5   Classification of quadrics   195
Solutions and hints for exercises   201

Index of notation   207

General index   208

EDITOR'S FOREWORD

Wide concern has been expressed in tertiary education about the difficulties experienced by students during their first year of an undergraduate course containing a substantial component of mathematics. These difficulties have a number of underlying causes, including the change of emphasis from an algorithmic approach at school to a more rigorous and abstract approach in undergraduate studies, the greater expectation of independent study, and the increased pace at which material is presented. The books in this series are intended to be sensitive to these problems.

Each book is a carefully selected, short, introductory text on a key area of the first-year syllabus; the areas are complementary and largely self-contained. Throughout, the pace of development is gentle, sympathetic and carefully motivated. Clear and detailed explanations are provided, and important concepts and results are stressed.

As mathematics is a practical subject which is best learned by doing it, rather than watching or reading about someone else doing it, a particular effort has been made to include a plentiful supply of worked examples, together with appropriate exercises, ranging in difficulty from the straightforward to the challenging.

When one goes fellwalking, the most breathtaking views require some expenditure of effort in order to gain access to them: nevertheless, the peak is more likely to be reached if a gentle and interesting route is chosen. The mathematical peaks attainable in these books are every bit as exhilarating, the paths are as gentle as we could find, and the interest and expectation are maintained throughout to prevent the spirits from flagging on the journey.

Lancaster, 1987 David A. Towers
Consultant Editor

PREFACE

'What! Not another linear algebra book!', I hear the reader cry—and with some justification. It is undoubtedly true that there is a very wide selection of books available on this topic, largely because it is so widely taught and applied, and because of the intrinsic beauty of the subject. However, while many of these books are very suitable for higher-level courses, fewer have been written with the specific needs of first-year undergraduates in mind. This guide is intended to be sympathetic to the problems often encountered during the transition from school to university mathematics, and this has influenced both the choice of material and its presentation.

All new concepts are introduced gently and are fully illustrated by examples. An unusual feature is the inclusion of exercises at the end of each section. A mathematical concept can be properly understood only by using it, and so it is important to develop and to test your understanding by attempting exercises after each new idea or set of ideas has been introduced. Many of these exercises are routine and are designed simply to help you to digest the definitions and to gain confidence. Others are more demanding and you should not be discouraged if you are unable to do them: sometimes a 'trick' ('insight' and 'inspiration' are commonly used synonyms) is involved, and most of us are inspired only very rarely. Nevertheless, much can be gained from wrestling with a difficult problem, even if a complete solution is not obtained. Full solutions to almost all of the exercises are included at the end of each chapter in order to make the book more useful if you are studying alone (and, I hope, to allow you the satisfaction of judging your solution to be superior to mine!). Only a knowledge of the basic notation and elementary ideas of set theory are assumed.

Abstraction lends enormous power to mathematics, as we learn early in life when we abstract the concept of number by observing the property that sets of two objects, for instance, have in common. Unfortunately, abstract ideas are assimilated very slowly, and so the level of abstraction is increased gently as the book progresses. The first chapter looks at vectors in two or three dimensions. These may well have been encountered previously in applied mathematics or physics, but are developed fully here in a way which lends itself to generalisation later. Geometrical ideas are stressed, as they underlie much of linear algebra.

Chapters 2 and 3 develop basic matrix algebra and apply it to the study of systems of linear equations. In Chapter 4, determinants are introduced, and here difficult choices have had to be made. I have chosen to give a treatment which is rather dated, but which is nevertheless rigorous enough for those demanding thoroughness. In doing so I have rejected the elegant modern approaches because of t.ie level of sophistication. I have also rejected the compromise which introduces simplification by treating only the 3×3 case rigorously, because of its lack of thoroughness. It is true that the proofs included here are not central to modern linear algebra and that it is important only to acquire a certain facility with the manipulation of determinants. However, those who find the proofs difficult or not to their taste can acquire this facility simply by concentrating on the results and gaining experience with the examples and exercises. Those of a genuine mathematical curiosity who refuse to believe anything until it is demonstrated to their satisfaction will, I hope, gain some pleasure from the proofs.

Chapters 5 to 7 are concerned with the basic ideas of modern linear algebra: the notions of vector spaces and of structure-preserving maps between them. It is shown how the more abstract concepts here grow naturally out of the earlier material, and how the more powerful tools developed can be applied to obtain deeper results on matrices and on systems of linear equations. The final chapter gives another application, this time to the classification of conics and quadrics, and thereby emphasising once again the underlying geometry.

While great attention has been devoted to developing the material in this book slowly and carefully, it must be realised that nothing of any value in this world is achieved without some expenditure of effort. You should not expect to understand everything on first reading, and should not be discouraged if you find a particular proof or idea difficult at first. This is normal and does not in itself indicate lack of ability or of aptitude. Carry on as best you can and come back to it later, or, in the case of a proof, skip it altogether and concentrate on understanding the statement of the result. Rests are also important: ideas are often assimilated and inspiration acquired (in a manner not understood) during rest. But do at least some of the exercises; never assume that you have understood without putting it to the test.

This book is based on many years of teaching algebra to undergraduates, and on many hours of listening carefully to their problems. It has been influenced, inevitably, by the many excellent teachers and colleagues I have been fortunate to have, and I am happy to record my thanks to them. Any qualities which the book may have are due in very large part to them; the shortcomings are all my own. Finally I would like to thank Peter Oates at Macmillans whose idea this series was, and without whose encouragement (and patience when deadlines were missed!) this book would never have been written.

Lancaster, 1987   D. A. T.

1 VECTORS

1.1 FREE VECTORS AND POSITION VECTORS

There are many examples in physics of quantities for which it is necessary to specify both a length and a direction. These include force, velocity and displacement, for example. Such quantities are modelled in mathematical terms by the concept of a vector. As so often happens in mathematics, the abstracted mathematical idea is more widely applicable than might have been imagined from the specific examples which led to its creation. We will see later that a further abstraction of our ideas produces a concept of even greater utility and breadth of application, namely that of a vector space. For the time being, in this chapter we will restrict our applications to the study of lines and planes in solid geometry, but the vectors used will be no different from those employed in mechanics, in physics and elsewhere.

A *free vector* is simply a directed line-segment; where there is no risk of confusion we will normally omit the word 'free', and say simply 'vector'. We say that two (free) vectors are *equal* if they have the same length and the same direction (though not necessarily the same position in space). See Fig. 1.1. We will denote vectors by small letters in boldface, such as **v**, or by $\overrightarrow{AB}$, where this means the line segment joining A and B and directed from A to B.

We denote the set of all vectors in space by $\mathbb{R}^3$; likewise, the set of all vectors in a single plane will be denoted by $\mathbb{R}^2$.

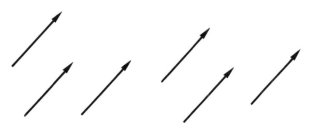

Fig. 1.1 **Equal free vectors**

The *sum* of two vectors **a** and **b** is defined to be the vector **a** + **b** represented by the diagonal $\overrightarrow{OC}$ of a parallelogram of which two adjacent sides $\overrightarrow{OA}$ and $\overrightarrow{OB}$ represent **a**, **b** respectively (Fig. 1.2).

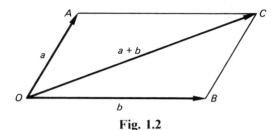

Fig. 1.2

This is usually called the *parallelogram law* of addition of vectors. It is equivalent to the so-called *triangle law* of addition: if $\overrightarrow{OA}$ represents **a** and $\overrightarrow{AB}$ represents **b**, then $\overrightarrow{OB}$ represents **a** + **b** (Fig. 1.3).

In using Fig. 1.3 it might help to recall one of the concrete situations being modelled by vectors: that of displacement. Imagine walking from O to A and then onwards from A to B; the net result of this journey is the same as that of walking directly from O to B.

Given a vector **a** we define $-\mathbf{a}$ to be the vector having the same length as **a** but opposite in direction to **a** (Fig. 1.4).

The *zero vector*, **0**, is the vector with length 0 and any direction which it happens to be convenient to give it. It does not really matter which direction we assign since the length is zero, but we must remember that the vector **0** is logically different from the real number 0.

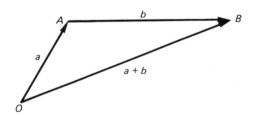

Fig. 1.3

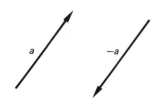

Fig. 1.4

Then vector addition satisfies the following laws:

V1 $(\mathbf{a} + \mathbf{b}) + \mathbf{c} = \mathbf{a} + (\mathbf{b} + \mathbf{c})$ for all $\mathbf{a}, \mathbf{b}, \mathbf{c} \in \mathbb{R}^3$.

(Unless otherwise stated, all of the results we assert for $\mathbb{R}^3$ will apply equally to vectors in $\mathbb{R}^2$, and we will not remark further upon this.) This is the *associative law* of addition of vectors, and it can be easily verified by considering Fig. 1.5.

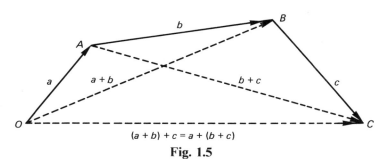

Fig. 1.5

Let $\overrightarrow{OA} = \mathbf{a}$, $\overrightarrow{AB} = \mathbf{b}$, $\overrightarrow{BC} = \mathbf{c}$. Then applying the triangle law of addition first to triangle OAB shows that $\overrightarrow{OB} = \mathbf{a} + \mathbf{b}$; and applying it to triangle OBC gives $\overrightarrow{OC} = (\mathbf{a} + \mathbf{b}) + \mathbf{c}$. Likewise, applying the triangle law of addition first to the triangle ABC shows that $\overrightarrow{AC} = \mathbf{b} + \mathbf{c}$; then application to triangle OAC gives $\overrightarrow{OC} = \mathbf{a} + (\mathbf{b} + \mathbf{c})$. Thus $\overrightarrow{OC} = (\mathbf{a} + \mathbf{b}) + \mathbf{c} = \mathbf{a} + (\mathbf{b} + \mathbf{c})$ as claimed.

V2 $\mathbf{a} + \mathbf{0} = \mathbf{0} + \mathbf{a} = \mathbf{a}$ for all $\mathbf{a} \in \mathbb{R}^3$.

This is saying that the zero vector acts as a *neutral* or *identity* element with respect to vector addition.

V3 $\mathbf{a} + (-\mathbf{a}) = (-\mathbf{a}) + \mathbf{a} = \mathbf{0}$ for all $\mathbf{a} \in \mathbb{R}^3$.

We say that $-\mathbf{a}$ is an additive *inverse* element for the vector $\mathbf{a}$.

V4 $\mathbf{a} + \mathbf{b} = \mathbf{b} + \mathbf{a}$ for all $\mathbf{a}, \mathbf{b} \in \mathbb{R}^3$.

This is the *commutative law* of addition of vectors.

Some readers will recognise V1 to V4 as being the axioms for an *abelian group*, so we could express them more succinctly by saying that the set of vectors in $\mathbb{R}^3$ forms an abelian group under vector addition. (If you have not come across the concept of an abelian group yet then simply ignore this remark!) Laws V1 to V4 imply that when forming arbitrary finite sums of vectors we can omit brackets and ignore order: the same vector results however we bracket the sum and in whatever order we write the vectors. In particular, we can form multiples of a given vector:

$$n\mathbf{a} = \mathbf{a} + \mathbf{a} + \ldots + \mathbf{a} \quad (n \text{ terms in the sum}) \quad \text{for all } n \in \mathbb{Z}^+.$$

We will denote the *length* of a vector **a** by $|\mathbf{a}|$ (and of $\overrightarrow{AB}$ by $|\overrightarrow{AB}|$). So $n\mathbf{a}$ is a vector in the same direction as **a** but with length $n|\mathbf{a}|$. It seems natural to define $\alpha\mathbf{a}$, where α is any positive real number to be the vector with the same direction as **a** and with length $\alpha|\mathbf{a}|$. Similarly, if α is a negative real number then we let $\alpha\mathbf{a}$ be the vector with the opposite direction to that of **a** and with length $-\alpha|\mathbf{a}|$. Finally we put $0\mathbf{a} = \mathbf{0}$. We now have a way of multiplying any vector in $\mathbb{R}^3$ by any real number. This multiplication is easily seen to satisfy the following laws:

V5 $\alpha(\mathbf{a} + \mathbf{b}) = \alpha\mathbf{a} + \alpha\mathbf{b}$ for all $\alpha \in \mathbb{R}$, and all $\mathbf{a}, \mathbf{b} \in \mathbb{R}^3$.

V6 $(\alpha + \beta)\mathbf{a} = \alpha\mathbf{a} + \beta\mathbf{a}$ for all $\alpha, \beta \in \mathbb{R}$, and all $\mathbf{a} \in \mathbb{R}^3$.

V7 $(\alpha\beta)\mathbf{a} = \alpha(\beta\mathbf{a})$ for all $\alpha, \beta \in \mathbb{R}$, and all $\mathbf{a} \in \mathbb{R}^3$.

V8 $1\mathbf{a} = \mathbf{a}$ for all $\mathbf{a} \in \mathbb{R}^3$.

V9 $0\mathbf{a} = \mathbf{0}$ for all $\mathbf{a} \in \mathbb{R}^3$.

We will encounter properties V1 to V9 again later where they will be used as the axioms for an abstract structure known as a *vector space*, but more of that later. We usually write $\alpha\mathbf{a} + (-\beta)\mathbf{b}$, where $\beta > 0$, as $\alpha\mathbf{a} - \beta\mathbf{b}$.

We have seen that, given any (free) vector there are infinitely many other (free) vectors which are equal to it. We often wish to pick a particular one of this class of (free) vectors. We do this by picking a fixed point O of space as the origin; the vector $\overrightarrow{OP}$ drawn from O to the point P is then called the *position vector* of P relative to O. Position and (free) vectors are powerful tools for proving results in geometry; often a result which requires a lengthy or complex proof using coordinate geometry succumbs far more readily to the vector approach. We will give an example.

Example 1.1

Prove that the lines joining the mid-points of the sides of any quadrilateral form a parallelogram.

Solution   Our first priority is to draw a helpful diagram (taking care not to make the quadrilateral look too 'special'—like a rectangle!) Let the quadrilateral (Fig. 1.6) have vertices A, B, C, D and label the mid-points by E, F, G, H as shown. Choose a point O as origin, and, relative to O, let the position vector of A be **a**, of B be **b**, and so on. Then

$$\overrightarrow{DA} = \mathbf{a} - \mathbf{d}, \quad \text{so} \quad \overrightarrow{HA} = \tfrac{1}{2}(\mathbf{a} - \mathbf{d});$$

and

$$\overrightarrow{AB} = \mathbf{b} - \mathbf{a}, \quad \text{so} \quad \overrightarrow{AE} = \tfrac{1}{2}(\mathbf{b} - \mathbf{a}).$$

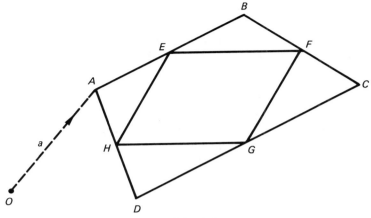

Fig. 1.6

Therefore, $\quad \vec{HE} = \vec{HA} + \vec{AE} = \frac{1}{2}(\mathbf{a} - \mathbf{d}) + \frac{1}{2}(\mathbf{b} - \mathbf{a}) = \frac{1}{2}(\mathbf{b} - \mathbf{d})$.

Also, $\quad \vec{DC} = \mathbf{c} - \mathbf{d}, \quad$ so $\quad \vec{GC} = \frac{1}{2}(\mathbf{c} - \mathbf{d})$;

and $\quad \vec{CB} = \mathbf{b} - \mathbf{c}, \quad$ so $\quad \vec{CF} = \frac{1}{2}(\mathbf{b} - \mathbf{c})$.

Therefore, $\quad \vec{GF} = \vec{GC} + \vec{CF} = \frac{1}{2}(\mathbf{c} - \mathbf{d}) + \frac{1}{2}(\mathbf{b} - \mathbf{c}) = \frac{1}{2}(\mathbf{b} - \mathbf{d})$.

We have shown that $\vec{HE} = \vec{GF}$ (as free vectors); in other words the sides EH and FG are equal in length and parallel to one another. It follows from this fact alone that $EFGH$ is a parallelogram, but it is just as easy to show that $\vec{EF} = \vec{HG}$. Why not try it?

EXERCISES 1.1

1 Let A, B be points with position vectors $\mathbf{a}, \mathbf{b}$ respectively relative to a fixed origin O.

  (a) Find $\vec{AB}$ in terms of $\mathbf{a}$ and $\mathbf{b}$.
  (b) If P is the mid-point of AB, find $\vec{AP}$.
  (c) Show that $\vec{OP} = \frac{1}{2}(\mathbf{a} + \mathbf{b})$.
  (d) If Q divides AB in the ratio $r:s$, find $\vec{AQ}$.
  (e) Find $\vec{OQ}$.

2 Prove that the line joining the mid-points of the two sides of a triangle is parallel to the third side and has half the length.

3 Prove that the diagonals of a parallelogram bisect each other.

4 Complete example 1.1 by showing that $\vec{EF} = \vec{HG}$.

1.2 BASES AND COORDINATES

Here we start to consider what is involved in setting up a coordinate system in space. Let $\{v_1, \ldots, v_n\}$ be a given set of vectors. If we can write the vector v in the form $v = \lambda_1 v_1 + \ldots + \lambda_n v_n$ for some $\lambda_1, \ldots, \lambda_n \in \mathbb{R}$, we say that v is a *linear combination* of $v_1, \ldots, v_n$.

The sequence $v_1, \ldots, v_n$ of vectors is *linearly dependent* if one of them can be written as a linear combination of the rest; otherwise it is said to be *linearly independent*. When faced with the problem of showing that a particular sequence of vectors is linearly independent we may often find it useful to have the equivalent formulation of this definition which is given in the following lemma.

LEMMA 1.2.1

(a) $v_1, \ldots, v_n$ are linearly dependent if and only if $\exists \lambda_1, \ldots, \lambda_n \in \mathbb{R}$, not all zero, such that $\lambda_1 v_1 + \ldots + \lambda_n v_n = 0$.

(b) $v_1, \ldots, v_n$ are linearly independent if and only if
$$(\lambda_1 v_1 + \ldots + \lambda_n v_n = 0 \quad \Rightarrow \quad \lambda_1 = \ldots = \lambda_n = 0).$$

Proof

(a) ($\Rightarrow$) Let $v_1, \ldots, v_n$ be linearly dependent. Then one of them, v_i say, can be written as
$$v_i = \lambda_1 v_1 + \ldots + \lambda_{i-1} v_{i-1} + \lambda_{i+1} v_{i+1} + \ldots + \lambda_n v_n.$$
Hence, by taking the v_i over to the other side of the equation, we get
$$\lambda_1 v_1 + \ldots + \lambda_{i-1} v_i + \lambda_i v_i + \lambda_{i+1} v_{i+1} + \ldots + \lambda_n v_n = 0$$
where $\lambda_i = -1 \neq 0$.

($\Leftarrow$) Suppose that $\lambda_1 v_1 + \ldots + \lambda_n v_n = 0$ where $\lambda_i \neq 0$. Then this can be rearranged to give
$$v_i = -\lambda_i^{-1}(\lambda_1 v_1 + \ldots + \lambda_{i-1} v_{i-1} + \lambda_{i+1} v_{i+1} + \ldots + \lambda_n v_n),$$
so that v_i is a linear combination of the rest, and $v_1, \ldots, v_n$ are linearly dependent.

(b) This is simply the contrapositive of (a).

Note: It is clear that any sequence of vectors containing 0 is linearly dependent. (Simply choose the coefficient of 0 to be non-zero and all of the other coefficients to be zero.)

Next we will consider what linear dependence means geometrically.

(a) Suppose first that v_1 and v_2 are linearly dependent. Then $v_1 = \lambda v_2$ for some $\lambda \in \mathbb{R}$. A moment's thought should be sufficient to realise that this means v_1 and v_2 are in the same or opposite direction and $|v_1| = |\lambda||v_2|$. For example,

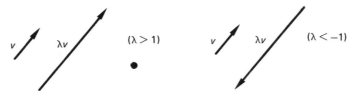

(b) Suppose next that v_1, v_2 and v_3 are linearly dependent. Then one of them, v_1 say, can be written as a linear combination of the other two. Hence $v_1 = \lambda v_2 + \mu v_3$, say, for some $\lambda, \mu \in \mathbb{R}$. This means that v_1, v_2 and v_3 lie in the same plane, as can be seen from the parallelogram law of addition of vectors (Fig. 1.7). Conversely, any three vectors lying in the same plane must be linearly dependent.

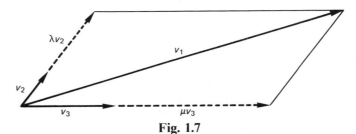

Fig. 1.7

(c) Similarly, any four vectors lying in three-dimensional space are linearly dependent. For, if there are three of them which are linearly independent, then these three do not lie in the same plane and the fourth can be written as a linear combination of them (Fig. 1.8).

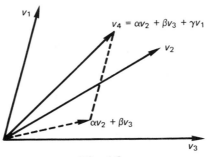

Fig. 1.8

So let e_1, e_2 and e_3 be three linearly independent vectors, and let v be any vector. Then v can be written as a linear combination of e_1, e_2 and e_3, thus

$$v = \lambda_1 e_1 + \lambda_2 e_2 + \lambda_3 e_3. \tag{1}$$

Moreover, this expression is unique. For, suppose that

$$\lambda_1 e_1 + \lambda_2 e_2 + \lambda_3 e_3 = \mu_1 e_1 + \mu_2 e_2 + \mu_3 e_3.$$

Then
$$(\lambda_1 - \mu_1)e_1 + (\lambda_2 - \mu_2)e_2 + (\lambda_3 - \mu_3)e_3 = 0.$$

Since e_1, e_2 and e_3 are linearly independent, we have $\lambda_1 - \mu_1 = 0, \lambda_2 - \mu_2 = 0, \lambda_3 - \mu_3 = 0$; that is, $\lambda_i = \mu_i$ for $i = 1, 2, 3$.

We call such a system of three linearly independent vectors a *basis* for $\mathbb{R}^3$. Similarly, a system of two linearly independent vectors is a basis for $\mathbb{R}^2$. We say that $\mathbb{R}^3$ is *three-dimensional* (and that $\mathbb{R}^2$ is *two-dimensional*). Thus, the dimension is the number of vectors in a basis for the space. The numbers $\lambda_1, \lambda_2, \lambda_3$ in equation (1) are called the *coordinates* of v relative to the basis e_1, e_2, e_3.

We can now set up a coordinate system in space. Pick a fixed point O as origin and take three linearly independent vectors e_1, e_2 and e_3. If P is any point in space then its position vector relative to O can be expressed uniquely as a linear combination of the e_is, namely $\overrightarrow{OP} = \lambda_1 e_1 + \lambda_2 e_2 + \lambda_3 e_3$. Thus, $\overrightarrow{OP}$, and hence P itself, is uniquely specified by its coordinates $(\lambda_1, \lambda_2, \lambda_3)$ relative to the given basis.

Note: A basis is an *ordered* sequence of vectors; changing the order of the vectors in a basis will change the order of the coordinates of a particular point, and so the resulting sequence must be considered to be a different basis. Also, it is implicit in the definition of basis that it must be a sequence of *non-zero* vectors.

Example 1.2

The vectors x, y and z have coordinates $(1, 0, 0)$, $(1, 1, 0)$ and $(1, 1, 1)$ relative to a basis for $\mathbb{R}^3$. Show that x, y and z also form a basis for $\mathbb{R}^3$, and express $(1, 3, 2)$ in terms of this new basis (that is, as a linear combination of x, y and z).

Solution We need to show that x, y and z are linearly independent, and we make use of Lemma 1.2.1(b). Now

$$\alpha(1, 0, 0) + \beta(1, 1, 0) + \gamma(1, 1, 1) = (0, 0, 0)$$
$$\Rightarrow \alpha + \beta + \gamma = 0, \quad \beta + \gamma = 0, \quad \gamma = 0$$
$$\Rightarrow \alpha = \beta = \gamma = 0.$$

Hence x, y and z are linearly independent.

Also,
$$(1, 3, 2) = \alpha(1, 0, 0) + \beta(1, 1, 0) + \gamma(1, 1, 1)$$

$$\Rightarrow 1 = \alpha + \beta + \gamma, \quad 3 = \beta + \gamma, \quad 2 = \gamma$$
$$\Rightarrow \gamma = 2, \quad \beta = 1, \quad \alpha = -2.$$

Therefore $\quad (1, 3, 2) = -2(1, 0, 0) + (1, 1, 0) + 2(1, 1, 1).$

EXERCISES 1.2

1 Show that the vectors $(1, 1, 1)$, $(1, 2, 3)$ and $(1, -1, -3)$ are linearly dependent.

2 Express each of the vectors in question 1 as a linear combination of the other two.

3 Show that the vectors $(1, -2, 0)$, $(0, 1, 4)$ and $(0, -1, -3)$ are linearly independent, and express $(-1, 2, 3)$ as a linear combination of them.

4 Show that the vectors $(-1, 1, 2)$, $(1, 1, 0)$ and $(3, -3, -6)$ are linearly dependent. Can each be expressed as a linear combination of the others?

5 Pick two of the vectors from question 4 which are linearly independent. Find a third vector which, together with these two, forms a basis for $\mathbb{R}^3$.

6 Show that the vectors $(1, 3)$ and $(2, 3)$ form a basis for $\mathbb{R}^2$.

1.3 SCALAR PRODUCT

So far we have no means of multiplying together two vectors. In this section and the next we will introduce two such means, one of which produces a *scalar* (that is, a real number) and the other a vector.

The *scalar product* of vectors **a** and **b**, whose directions are inclined at an angle θ, is the real number $|\mathbf{a}||\mathbf{b}|\cos\theta$, and is written $\mathbf{a} \cdot \mathbf{b}$*. Because of the dot this is sometimes referred to as the *dot product*.

Note: The angle θ is measured between the two 'pointed ends' of **a** and **b**, and $0 \leq \theta \leq \pi$.

Clearly (Fig. 1.9) $OP/OA = \cos\theta$ and so $|\mathbf{a}|\cos\theta = OA\cos\theta = OP =$ projection of $|\mathbf{a}|$ onto **b**.

Thus $\quad \mathbf{a} \cdot \mathbf{b} = |\mathbf{a}||\mathbf{b}|\cos\theta = |\mathbf{b}| \times$ (the projection of $|\mathbf{a}|$ onto **b**).

Similarly, $\quad \mathbf{a} \cdot \mathbf{b} = |\mathbf{a}| \times$ (the projection of $|\mathbf{b}|$ onto **a**).

The scalar product has the following properties:

1 $\mathbf{a} \cdot \mathbf{b} = \mathbf{b} \cdot \mathbf{a} \quad$ for all $\mathbf{a}, \mathbf{b} \in \mathbb{R}^3$.

---

* *Note:* This is often written as **a.b**. Here the dot has been set in the centre of the line to distinguish it from a decimal point and from a multiplication point.

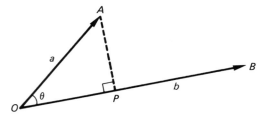

Fig. 1.9

2 $(\lambda\mathbf{a} + \mu\mathbf{b})\cdot\mathbf{c} = \lambda\mathbf{a}\cdot\mathbf{c} + \mu\mathbf{b}\cdot\mathbf{c}$ for all $\lambda, \mu \in \mathbb{R}$, and all $\mathbf{a}, \mathbf{b}, \mathbf{c} \in \mathbb{R}^3$.

3 If $\theta = 0$, then $\cos\theta = 1$, so that $\mathbf{a}\cdot\mathbf{a} = |\mathbf{a}|^2$. Thus $|\mathbf{a}| = \sqrt{\mathbf{a}\cdot\mathbf{a}}$.

4 Suppose that $\mathbf{a} \neq \mathbf{0}$, $\mathbf{b} \neq \mathbf{0}$. Then

$$\mathbf{a}\cdot\mathbf{b} = 0 \quad \Leftrightarrow \quad |\mathbf{a}||\mathbf{b}|\cos\theta = 0 \quad \Leftrightarrow \quad \cos\theta = 0 \quad \Leftrightarrow \quad \theta = \tfrac{1}{2}\pi.$$

When $\mathbf{a}\cdot\mathbf{b} = 0$ we say that $\mathbf{a}, \mathbf{b}$ are *orthogonal* (or *perpendicular*).

A *unit* vector is one with length 1. Clearly, if $\mathbf{a} \neq \mathbf{0}$ then $\mathbf{a}/|\mathbf{a}|$ is a unit vector. Let $\mathbf{i}, \mathbf{j}, \mathbf{k}$ be a set of three mutually orthogonal unit vectors. (By mutually orthogonal we mean that $\mathbf{i}\cdot\mathbf{j} = \mathbf{i}\cdot\mathbf{k} = \mathbf{j}\cdot\mathbf{k} = 0$.) Then these vectors form a basis for $\mathbb{R}^3$. For, suppose that $\alpha\mathbf{i} + \beta\mathbf{j} + \gamma\mathbf{k} = \mathbf{0}$; then, taking the scalar product of each side with $\mathbf{i}$ and using property 2 above,

$$0 = \mathbf{0}\cdot\mathbf{i} = \alpha\mathbf{i}\cdot\mathbf{i} + \beta\mathbf{j}\cdot\mathbf{i} + \gamma\mathbf{k}\cdot\mathbf{i} = \alpha|\mathbf{i}|^2 = \alpha.$$

Similarly, by taking the scalar product of each side with $\mathbf{j}$, and then with $\mathbf{k}$, we find that $\beta = \gamma = 0$. We have thus justified our assertion that this is a basis; such a basis is called a *rectangular coordinate system for* $\mathbb{R}^3$. There are essentially two possible orientations for such a system (Fig. 1.10). In the right-handed system, if you imagine a right-handed screw being turned in a direction from $\mathbf{i}$ to $\mathbf{j}$ the point of the screw would move in the direction of $\mathbf{k}$.

From now on, whenever we write coordinates for a vector without comment, they will be relative to a right-handed rectangular coordinate system i, j, k.

right-handed system   left-handed system

Fig. 1.10

We seek next an expression for the scalar product in terms of coordinates.

Let $\mathbf{a} = (a_1, a_2, a_3)$ $\quad (= a_1\mathbf{i} + a_2\mathbf{j} + a_3\mathbf{k}$, remember)

$\mathbf{b} = (b_1, b_2, b_3)$.

Then $\mathbf{a} \cdot \mathbf{b} = (a_1\mathbf{i} + a_2\mathbf{j} + a_3\mathbf{k}) \cdot (b_1\mathbf{i} + b_2\mathbf{j} + b_3\mathbf{k})$

$= a_1 b_1 + a_2 b_2 + a_3 b_3,$

using property 2 of scalar product and the fact that $\mathbf{i} \cdot \mathbf{j} = \mathbf{i} \cdot \mathbf{k} = \mathbf{j} \cdot \mathbf{k} = 0$ and $\mathbf{i} \cdot \mathbf{i} = \mathbf{j} \cdot \mathbf{j} = \mathbf{k} \cdot \mathbf{k} = 1$.

Examples 1.3

1. Find $\cos \theta$, where θ is the angle between $\mathbf{a} = (1, 2, 4)$ and $\mathbf{b} = (0, 1, 1)$.

   Solution $\mathbf{a} \cdot \mathbf{b} = 1 \times 0 + 2 \times 1 + 4 \times 1 = 6;$

   $|\mathbf{a}| = \sqrt{\mathbf{a} \cdot \mathbf{a}} = \sqrt{1^2 + 2^2 + 4^2} = \sqrt{21}; \quad |\mathbf{b}| = \sqrt{\mathbf{b} \cdot \mathbf{b}} = \sqrt{0^2 + 1^2 + 1^2} = \sqrt{2}$

   $$\cos \theta = \frac{\mathbf{a} \cdot \mathbf{b}}{|\mathbf{a}||\mathbf{b}|} = \frac{6}{\sqrt{42}}.$$

2. Find a unit vector $\mathbf{u}$ orthogonal to $(1, 1, 1)$ and $(1, 0, 0)$.

   Solution
   First we seek a vector $\mathbf{a} = (x, y, z)$ which is orthogonal to the two given vectors; we want, therefore, the scalar product of each of them with $\mathbf{a}$ to be zero. Thus

   $$0 = (x, y, z) \cdot (1, 1, 1) = x + y + z$$

   and $$0 = (x, y, z) \cdot (1, 0, 0) = x.$$

   Solving, we have $\quad x = 0, \; y + z = 0; \quad$ that is, $\quad x = 0, \; y = -z.$

   Therefore the vector $\mathbf{a} = (0, -z, z)$ is orthogonal to the two given vectors. Now $|\mathbf{a}| = \sqrt{z^2 + z^2} = \sqrt{2}z$, so choosing $z = 1/\sqrt{2}$ gives the required unit vector,

   $$\mathbf{u} = (0, -1/\sqrt{2}, 1/\sqrt{2}).$$

   It is clear that $-\mathbf{u}$ is an equally valid answer.

EXERCISES 1.3

1 Let $\mathbf{v} = (1, 1)$, $\mathbf{w} = (1, 2)$. Find (a) $\mathbf{v} \cdot \mathbf{w}$, (b) $\cos \theta$, where θ is the angle between $\mathbf{v}$ and $\mathbf{w}$.

2 Let $\mathbf{v} = (1, 1, 1)$, $\mathbf{w} = (1, 2, 0)$. Find (a) $\mathbf{v} \cdot \mathbf{w}$, (b) $\cos \theta$, where θ is the angle between $\mathbf{v}$ and $\mathbf{w}$.

3 Find the lengths of the vectors in questions 1 and 2 above.

4 Find a vector which is orthogonal to $(1, 2, 4)$ and $(2, 1, -1)$.

5 Find a vector **c** which is orthogonal to $(1, 3, 1) = \mathbf{a}$ and to $(2, 1, 1) = \mathbf{b}$, and verify that $\mathbf{a}, \mathbf{b}, \mathbf{c}$ is a basis for $\mathbb{R}^3$.

6 By expanding $(\mathbf{a} + \mathbf{b}) \cdot (\mathbf{a} + \mathbf{b})$, show that, for a triangle with sides a, b, c and corresponding angles α, β, γ (that is, angle α is opposite side a, etc.), $a^2 + b^2 - c^2 = 2ab \cos \gamma$.

7 Show that the vectors $\mathbf{x} - \mathbf{y}$ and $\mathbf{x} + \mathbf{y}$ are orthogonal if and only if $|\mathbf{x}| = |\mathbf{y}|$. Deduce that the diagonals of a parallelogram are orthogonal if and only if it is a rhombus.

8 Using the expression for the scalar product in terms of coordinates, check property 2 (p. 10) for scalar products.

1.4 VECTOR PRODUCT

The *vector product* of two vectors **a** and **b**, whose directions are inclined at an angle θ, is the vector, denoted by $\mathbf{a} \times \mathbf{b}$, whose length is $|\mathbf{a}||\mathbf{b}| \sin \theta$ and whose direction is orthogonal to **a** and to **b**, being positive relative to a rotation from **a** to **b**. The last phrase in the preceeding sentence means that the direction of $\mathbf{a} \times \mathbf{b}$ is the same as that in which the point of a right-handed screw would move if turned from **a** to **b** (cf. a right-handed rectangular coordinate system) (Fig. 1.11). It is straightforward to check that the following properties are satisfied.

1    $\mathbf{a} \times \mathbf{b} = -\mathbf{b} \times \mathbf{a}$ for all $\mathbf{a}, \mathbf{b} \in \mathbb{R}^3$.

2    $(\lambda \mathbf{a} + \mu \mathbf{b}) \times \mathbf{c} = \lambda \mathbf{a} \times \mathbf{c} + \mu \mathbf{b} \times \mathbf{c}$ for all $\lambda, \mu \in \mathbb{R}$, and all $\mathbf{a}, \mathbf{b}, \mathbf{c} \in \mathbb{R}^3$.

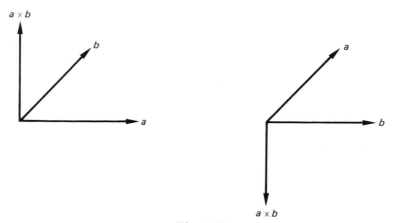

Fig. 1.11

3 If $\theta = 0$, then $\sin \theta = 0$, so that $\mathbf{a} \times \mathbf{a} = \mathbf{0}$.

4 If $\mathbf{i}, \mathbf{j}, \mathbf{k}$ is a right-handed rectangular coordinate system, then

$$\mathbf{i} \times \mathbf{j} = \mathbf{k}, \mathbf{j} \times \mathbf{k} = \mathbf{i}, \mathbf{k} \times \mathbf{i} = \mathbf{j}.$$

These equations are easy to remember because of their cyclical nature:

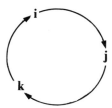

Notice that going around the cycle in an anticlockwise direction produces negative signs: $\mathbf{j} \times \mathbf{i} = -\mathbf{k}, \mathbf{k} \times \mathbf{j} = -\mathbf{i}, \mathbf{i} \times \mathbf{k} = -\mathbf{j}$.

As with scalar product we can find a formula for the vector product of two vectors whose coordinates are known relative to a right-handed rectangular coordinate system. Let

$$\mathbf{a} = (a_1, a_2, a_3), \quad \mathbf{b} = (b_1, b_2, b_3).$$

Then
$$\mathbf{a} \times \mathbf{b} = (a_1 \mathbf{i} + a_2 \mathbf{j} + a_3 \mathbf{k}) \times (b_1 \mathbf{i} + b_2 \mathbf{j} + b_3 \mathbf{k})$$
$$= a_1 b_2 \mathbf{k} - a_1 b_3 \mathbf{j} - a_2 b_1 \mathbf{k} + a_2 b_3 \mathbf{i} + a_3 b_1 \mathbf{j} - a_3 b_2 \mathbf{i}$$
$$= (a_2 b_3 - a_3 b_2)\mathbf{i} + (a_3 b_1 - a_1 b_3)\mathbf{j} + (a_1 b_2 - a_2 b_1)\mathbf{k}.$$

Thus $\mathbf{a} \times \mathbf{b} = (a_2 b_3 - a_3 b_2, a_3 b_1 - a_1 b_3, a_1 b_2 - a_2 b_1).$

Example 1.4

Let us try example 1.3.2 above again. That is, find a unit vector orthogonal to $(1, 1, 1)$ and $(1, 0, 0)$.

Solution A vector orthogonal to $(1, 1, 1)$ and to $(1, 0, 0)$ is

$$(1, 1, 1) \times (1, 0, 0) = (1 \times 0 - 1 \times 0, 1 \times 1 - 1 \times 0, 1 \times 0 - 1 \times 1)$$
$$= (0, 1, -1).$$

From this we can produce a unit vector exactly as before.

EXERCISES 1.4

1 Let $\mathbf{v} = (1, 1, 0)$, $\mathbf{w} = (2, 5, 0)$. Find (a) $\mathbf{v} \times \mathbf{w}$, (b) $|\mathbf{v}|, |\mathbf{w}|$ and $|\mathbf{v} \times \mathbf{w}|$, (c) $\sin \theta$, where θ is the angle between $\mathbf{v}$ and $\mathbf{w}$.

2 Let $\mathbf{v} = (1, 1, -1)$, $\mathbf{w} = (-1, 2, 0)$. Find (a) $\mathbf{v} \times \mathbf{w}$, (b) $|\mathbf{v}|, |\mathbf{w}|$ and $|\mathbf{v} \times \mathbf{w}|$, (c) $\sin \theta$, where θ is the angle between $\mathbf{v}$ and $\mathbf{w}$.

3 Find a unit vector which is orthogonal to $(0, -1, 2)$ and $(3, 1, 1)$.

4 By expanding $\mathbf{a} \times (\mathbf{a} + \mathbf{b} + \mathbf{c})$, show that, for a triangle with sides a, b, c and corresponding angles α, β, γ,

$$\frac{a}{\sin \alpha} = \frac{b}{\sin \beta} = \frac{c}{\sin \gamma}.$$

5 Using the expression for vector product in terms of coordinates, check property 2 (p. 12) for vector products.

6 Show that, if $\mathbf{a}, \mathbf{b}, \mathbf{c} \in \mathbb{R}^3$, then

$$(\mathbf{a} \times \mathbf{b}) \times \mathbf{c} + (\mathbf{b} \times \mathbf{c}) \times \mathbf{a} + (\mathbf{c} \times \mathbf{a}) \times \mathbf{b} = \mathbf{0}.$$

1.5 GEOMETRY OF LINES AND PLANES IN $\mathbb{R}^3$

There are a number of ways of specifying a particular line or plane in $\mathbb{R}^3$, and so the equation of such a line or plane can be expressed in a number of equivalent forms. Here we will consider some of those forms; which one is useful in a given context will depend upon the information available in that context.

(a) The equation of a line through a point with position vector a parallel to b.

First we can specify a line by giving a vector which lies along it, together with the position vector of a point lying on the line. Consider the situation shown in Fig. 1.12. We are given the position vector, $\mathbf{a}$, relative to a fixed origin O, of a point A which lies on the line, and a vector $\mathbf{b}$ which lies in the same direction as the line. Our task is to write the position vector, $\mathbf{r}$, of an arbitrary point P on the line in terms of $\mathbf{a}$ and $\mathbf{b}$. This we can do by applying

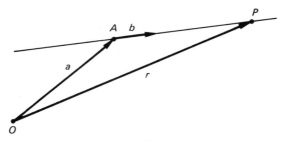

Fig. 1.12

the triangle law of addition to the triangle OAP. Then $\mathbf{r} = \vec{OA} + \vec{AP}$. Now $\vec{AP} = \lambda \mathbf{b}$ for some $\lambda \in \mathbb{R}$, and so

$$\mathbf{r} = \mathbf{a} + \lambda \mathbf{b} \qquad (\lambda \in \mathbb{R}).$$

This is the *vector form* of the equation. We can write it in terms of coordinates, or in *Cartesian form*, as follows.

Let $\mathbf{r} = (x, y, z)$, $\mathbf{a} = (a_1, a_2, a_3)$, $\mathbf{b} = (b_1, b_2, b_3)$. Substituting into the vector form of the equation gives

$$(x, y, z) = (a_1 + \lambda b_1, a_2 + \lambda b_2, a_3 + \lambda b_3).$$

Hence $\qquad x = a_1 + \lambda b_1,\ y = a_2 + \lambda b_2,\ z = a_3 + \lambda b_3.$

If $b_1, b_2, b_3 \neq 0$ then eliminating λ from these equations yields

$$\frac{x - a_1}{b_1} = \frac{y - a_2}{b_2} = \frac{z - a_3}{b_3} \qquad (= \lambda).$$

Examples 1.5.1

1. Find the equation(s), in Cartesian form, of the line through $(-2, 0, 5)$ parallel to $(1, 2, -3)$.

   Solution
   Substituting directly into the final formula gives the equations to be

   $$\frac{x+2}{1} = \frac{y}{2} = \frac{z-5}{-3}; \quad \text{that is,} \quad 3y = 6x + 12 = 10 - 2z.$$

2. Find the equation(s), in Cartesian form, of the line through $(1, 2, 3)$ parallel to $(-2, 0, 5)$.

   Solution Here we need to be a little careful as $b_2 = 0$. We therefore use the first of the Cartesian forms of the equations, namely

   $$x = 1 - 2\lambda, \qquad y = 2 + 0\lambda, \qquad z = 3 + 5\lambda.$$

   Eliminating λ gives $y = 2$ and $\dfrac{1-x}{2} = \dfrac{z-3}{5}$, so the equations are

   $$y = 2, \qquad 5x + 2z = 11$$

   Note: Both of these equations are needed to describe the line; each equation on its own describes a plane, as we shall see later, and the required line is the intersection of these two planes.

GUIDE TO LINEAR ALGEBRA

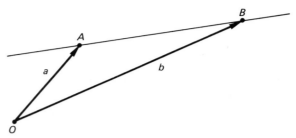

Fig. 1.13

(b) The equation of a line through two points

A line may also be uniquely specified by giving two points which lie on it. Suppose that the points A and B, with position vectors $\mathbf{a}$ and $\mathbf{b}$ respectively, lie on the line (Fig. 1.13). Then the vector $\overrightarrow{AB} = \mathbf{b} - \mathbf{a}$ lies along the line. We now have the information required in (a) above. By (a), the vector equation is

$$\mathbf{r} = \mathbf{a} + \lambda(\mathbf{b} - \mathbf{a}) = (1 - \lambda)\mathbf{a} + \lambda\mathbf{b} \qquad (\lambda \in \mathbb{R}).$$

In Cartesian form this will be

$$x = a_1 + \lambda(b_1 - a_1), \qquad y = a_2 + \lambda(b_2 - a_2), \qquad z = a_3 + \lambda(b_3 - a_3);$$

or

$$\frac{x - a_1}{b_1 - a_1} = \frac{y - a_2}{b_2 - a_2} = \frac{z - a_3}{b_3 - a_3}$$

if the denominators are non-zero.

Example 1.5.2

Prove that the medians of a triangle are concurrent. (A *median* of a triangle is a line joining a vertex to the mid-point of the opposite side. The point in which they meet is called the *centroid* of the triangle.)

Solution
Label the vertices of the triangle A, B, C and the mid-points of the sides D, E, F as shown in Fig. 1.14. Let G be the point of intersection of AD and BE, and let the position vectors, relative to a fixed origin O, of A be $\mathbf{a}$, of B be $\mathbf{b}$, and so on. Then, using exercise 1.1.1(c), the position vectors of F, D and E are $\frac{1}{2}(\mathbf{a} + \mathbf{b})$, $\frac{1}{2}(\mathbf{b} + \mathbf{c})$ and $\frac{1}{2}(\mathbf{a} + \mathbf{c})$ respectively. It follows from (b) above that any point on the line BE is of the form $(1 - \lambda)\mathbf{b} + \lambda(\frac{1}{2}(\mathbf{a} + \mathbf{c}))$. Similarly, any point on the line AD is of the form $(1 - \mu)\mathbf{a} + \mu(\frac{1}{2}(\mathbf{b} + \mathbf{c}))$. Since G lies on both of these lines, it must be possible to write it in each of these ways.

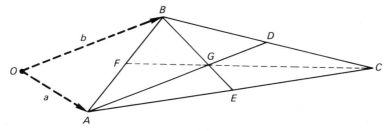

Fig. 1.14

Thus, at the point G,

$$(1-\lambda)\mathbf{b} + \tfrac{1}{2}\lambda\mathbf{a} + \tfrac{1}{2}\lambda\mathbf{c} = (1-\mu)\mathbf{a} + \tfrac{1}{2}\mu\mathbf{b} + \tfrac{1}{2}\mu\mathbf{c}. \qquad (1)$$

This will clearly be satisfied if $1-\lambda = \tfrac{1}{2}\mu$ and $\tfrac{1}{2}\lambda = 1-\mu$. Solving gives $\lambda = \tfrac{2}{3}$ and $\mu = \tfrac{2}{3}$. Substituting into either side of equation (1) shows that G is the point $\tfrac{1}{3}(\mathbf{a}+\mathbf{b}+\mathbf{c})$. It is straightforward to check that this also lies on the line CF. (Do it!)

(c) The equation of the plane through the origin and parallel to *a* and to *b*

Let the plane be that of this page, and suppose that the vectors $\overrightarrow{OA} = \mathbf{a}$ and $\overrightarrow{OB} = \mathbf{b}$ lie in this plane (Fig. 1.15). Then the position vector of a general point P in the plane is given by

$$\mathbf{r} = \overrightarrow{OP} = \lambda\mathbf{a} + \mu\mathbf{b} \qquad (\lambda, \mu \in \mathbb{R})$$

by the parallelogram law of addition.

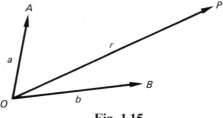

Fig. 1.15

(d) The equation of the plane through *c* parallel to *a* and to *b*

A general plane can be specified by giving two vectors which lie in the plane, thereby giving its direction, and the position vector of a point lying in the plane. Consider the situation shown in Fig. 1.16: vectors $\overrightarrow{CA} = \mathbf{a}$ and $\overrightarrow{CB} = \mathbf{b}$ lie in the plane, and the point C, with position vector relative to O equal to $\mathbf{c}$, lies also in the plane. Then $\overrightarrow{CP} = \lambda\mathbf{a} + \mu\mathbf{b}\,(\lambda, \mu \in \mathbb{R})$ as in (c) above. So, if P is a general point in the plane, its position vector relative to O is given by

$$\mathbf{r} = \overrightarrow{OP} = \overrightarrow{OC} + \overrightarrow{CP} = \mathbf{c} + \lambda\mathbf{a} + \mu\mathbf{b}.$$

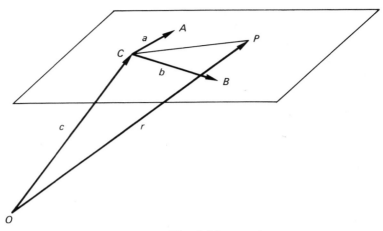

Fig. 1.16

(e) The equation of the plane through points a, b and c

We can also specify a plane uniquely by giving three non-collinear points which lie on it. Let A, B, C, with position vectors $\mathbf{a}, \mathbf{b}, \mathbf{c}$ respectively, lie on the plane, as shown in Fig. 1.17. Then the vectors $\overrightarrow{AB} = \mathbf{b} - \mathbf{a}$ and $\overrightarrow{AC} = \mathbf{c} - \mathbf{a}$ lie in the plane. Thus, if P is a general point in the plane, its position vector $\mathbf{r}$ relative to O is given by

$$\mathbf{r} = \mathbf{a} + \lambda(\mathbf{b} - \mathbf{a}) + \mu(\mathbf{c} - \mathbf{a}) \qquad (\lambda, \mu \in \mathbb{R}),$$

by (d) above. Hence

$$\mathbf{r} = (1 - \lambda - \mu)\mathbf{a} + \lambda\mathbf{b} + \mu\mathbf{c} \qquad (\lambda, \mu \in \mathbb{R}).$$

The vector equations obtained in (c), (d) and (e) above can, of course, be turned into Cartesian equations as in (a) and (b). We will simply illustrate with an example.

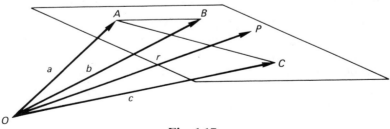

Fig. 1.17

Example 1.5.3

Find, in Cartesian form, the equation of the plane through $(1, 2, 3)$, $(1, 0, 0)$ and $(0, 1, 0)$.

Solution From (e) above, the vector equation is
$$(x, y, z) = (1 - \lambda - \mu)(1, 2, 3) + \lambda(1, 0, 0) + \mu(0, 1, 0).$$
Equating corresponding coordinates on each side we get
$$x = 1 - \lambda - \mu + \lambda = 1 - \mu \tag{1}$$
$$y = 2 - 2\lambda - 2\mu + \mu = 2 - 2\lambda - \mu \tag{2}$$
$$z = 3 - 3\lambda - 3\mu \tag{3}$$
Substituting (1) into (2) and (3) gives
$$y = 1 - 2\lambda + x \tag{4}$$
$$z = 3x - 3\lambda \tag{5}$$
$3 \times (4) - 2 \times (5)$: $3y - 2z = 3 - 6x + 3x$. The required equation is, therefore,
$$\underline{3x + 3y - 2z = 3.}$$

(f) The equation of a plane in terms of the normal to the plane

Let ON be the perpendicular from the origin to the plane (Fig. 1.18), and put $ON = p$. Let $\hat{\mathbf{n}}$ be a unit vector perpendicular to the plane in the direction ON. Then $\mathbf{r} \cdot \hat{\mathbf{n}}$ is the projection of OP onto ON, which is ON itself. Thus,
$$\mathbf{r} \cdot \hat{\mathbf{n}} = p,$$
which is the vector equation of the plane.

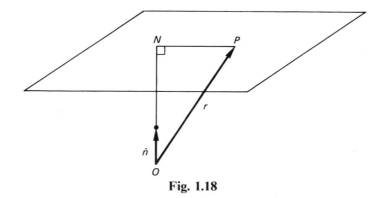

Fig. 1.18

Let $\hat{\mathbf{n}} = (a, b, c)$, $\mathbf{r} = (x, y, z)$. Then the equation becomes

$$ax + by + cz = p,$$

which is the Cartesian form of the equation.

Example 1.5.4

Consider the plane in the example above, namely $3x + 3y - 2z = 3$. This can be written as

$$(x, y, z) \cdot (3, 3, -2) = 3$$

which is in the form $\mathbf{p} \cdot \mathbf{n} = p'$, where $p' = 3$ and $\mathbf{n} = (3, 3, -2)$ is normal to the plane but is not a unit vector. Now

$$|\mathbf{n}| = \sqrt{3^2 + 3^2 + (-2)^2} = \sqrt{22}.$$

So the equation

$$(x, y, z) \cdot (3/\sqrt{22}, 3/\sqrt{22}, -2/\sqrt{22}) = 3/\sqrt{22}$$

is in the form $\mathbf{r} \cdot \hat{\mathbf{n}} = p$. It follows that the distance of this plane from the origin is $p = 3/\sqrt{22}$.

(g) The distance of a given plane from a given point, A

Let the plane have equation $\mathbf{r} \cdot \hat{\mathbf{n}} = p$, and let $\overrightarrow{OA} = \mathbf{a}$. Three possibilities arise, as illustrated in Fig. 1.19. In each case OM and AN are perpendicular to the plane. In (a) and (c), AD is perpendicular to OM (produced in the case of (c)); in (b) OD is perpendicular to AN. Consider each case separately:

(a) $AN = DM = OM - OD = p - \mathbf{a} \cdot \hat{\mathbf{n}}$;

(b) $AN = AD + DN = AD + OM = -\mathbf{a} \cdot \hat{\mathbf{n}} + p$;

(c) $AN = DM = OD - OM = \mathbf{a} \cdot \hat{\mathbf{n}} - p$.

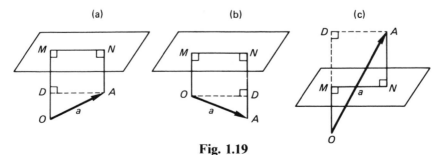

Fig. 1.19

Thus the distance of the point A from the plane is $|p - \mathbf{a} \cdot \hat{\mathbf{n}}|$. Furthermore, A is on the same side of the plane as the origin precisely when $p - \mathbf{a} \cdot \hat{\mathbf{n}}$ is positive.

Example 1.5.5

The distance of the point $(0, 0, 1)$ from the plane $3x + 3y - 2z = 3$ is
$$\frac{3}{\sqrt{22}} - (0, 0, 1) \cdot \frac{1}{\sqrt{22}}(3, 3, -2) = \frac{3}{\sqrt{22}} + \frac{2}{\sqrt{22}} = \frac{5}{\sqrt{22}}.$$
It is on the same side of the plane as the origin.

EXERCISES 1.5

1. Write down, in Cartesian form, the equation of the straight line through the point with position vector $(1, 2, 3)$ parallel to the direction given by $(1, 1, 0)$.

2. Find, in Cartesian form, the equation of the line through points $(1, 1, 1)$ and $(1, 1, 2)$.

3. Find, in Cartesian form, the equation of the plane through the points $(1, 0, 0)$, $(0, 1, 0)$ and $(0, 0, 1)$.

4. Find the equation of the plane which is perpendicular to the vector $(1, -1, 0)$ and which contains the point $(1, 2, 1)$. Find also the perpendicular distance of this plane from the origin.

5. Find a unit vector which is perpendicular to the plane containing the points $(0, 0, 0)$, $(1, 2, 3)$ and $(-4, 2, 2)$. Find also the perpendicular distance of this plane from the point $(2, 3, 4)$.

6. Let $A = (0, 0, 3)$, $B = (1, 1, 5)$, $C = (0, 3, 0)$ and $D = (-1, 5, 1)$ be four points in $\mathbb{R}^3$. Find:
   (a) the vector equation of the line through A, B and that of the line through C, D;
   (b) a unit vector $\hat{\mathbf{n}}$ orthogonal to both $\vec{AB}$ and $\vec{CD}$;
   (c) points P on AB and Q on CD such that $\vec{PQ}$ is orthogonal to both $\vec{AB}$ and $\vec{CD}$.
   Verify that $|\vec{PQ}| = \vec{AC} \cdot \hat{\mathbf{n}} = \vec{BD} \cdot \hat{\mathbf{n}}$.

7. Let OAB be a triangle, C be the mid-point of AB and D be the point on OB which divides it in the ratio $3:1$. The lines DC and OA meet at E. The position vectors of A and B relative to O are $\mathbf{a}$ and $\mathbf{b}$ respectively.

(a) Find the equation of the line CD in terms of **a** and **b**.
(b) Hence find $\overrightarrow{OE}$ in terms of **a**.
(c) Suppose now that $\overrightarrow{CD}$ and $\overrightarrow{OD}$ are orthogonal. Find $\cos(\widehat{AOB})$ in terms of $|\mathbf{a}|$ and $|\mathbf{b}|$.

8 Let $OABC$ be a parallelogram, P be the mid-point of OA, S be the mid-point of OC and R be the point of intersection of PB and AS. Find $PR:RB$.

9 Prove that the perpendicular bisectors of the sides of a triangle are concurrent. (The point of intersection is called the *circumcentre* of the triangle.)

10 Prove that the altitudes of a triangle are concurrent. (An *altitude* is a perpendicular from a vertex to the opposite side, possibly extended. The point of intersection is called the *orthocentre* of the triangle.)

11 Prove that the centroid, the circumcentre and the orthocentre of a triangle are collinear. (The line on which they all lie is called the *Euler line*.)

12 Complete the solution to example 1.5.2 by showing that G lies on CF.

SOLUTIONS AND HINTS FOR EXERCISES

Exercises 1.1

1 (a) $\mathbf{b} - \mathbf{a}$;   (b) $\frac{1}{2}(\mathbf{b} - \mathbf{a})$;
(c) $\overrightarrow{OP} = \overrightarrow{OA} + \overrightarrow{AP} = \mathbf{a} + \frac{1}{2}(\mathbf{b} - \mathbf{a}) = \frac{1}{2}(\mathbf{a} + \mathbf{b})$;   (d) $(r/(r+s))(\mathbf{b} - \mathbf{a})$;
(e) $\overrightarrow{OQ} = \overrightarrow{OA} + \overrightarrow{AQ} = (s/(r+s))\mathbf{a} + (r/(r+s))\mathbf{b}$.

2 (Fig. 1.20) Let A, B, C, D, E have position vectors $\mathbf{a}, \mathbf{b}, \mathbf{c}, \mathbf{d}, \mathbf{e}$ respectively relative to a fixed origin O. Then $\mathbf{d} = \frac{1}{2}(\mathbf{a} + \mathbf{c})$, $\mathbf{e} = \frac{1}{2}(\mathbf{c} + \mathbf{b})$ (as in question 1). Thus, $\overrightarrow{DE} = \mathbf{e} - \mathbf{d} = \frac{1}{2}(\mathbf{b} - \mathbf{a}) = \frac{1}{2}\overrightarrow{AB}$; whence result.

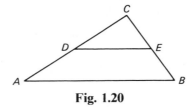

Fig. 1.20

3 (Fig. 1.21) Let P, Q be the mid-points of AC, BD respectively, and let A, B, C, D, P, Q have position vectors $\mathbf{a}, \mathbf{b}, \mathbf{c}, \mathbf{d}, \mathbf{p}, \mathbf{q}$ relative to a fixed origin O.
Then $\mathbf{p} = \frac{1}{2}(\mathbf{a} + \mathbf{c})$, $\mathbf{q} = \frac{1}{2}(\mathbf{b} + \mathbf{d})$.
Now $\overrightarrow{BC} = \overrightarrow{AD}$, so $\mathbf{c} - \mathbf{b} = \mathbf{d} - \mathbf{a}$. Rearranging this equation gives $\mathbf{a} + \mathbf{c} = \mathbf{b} + \mathbf{d}$, from which $\mathbf{p} = \mathbf{q}$, and hence $P = Q$.

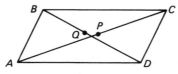

Fig. 1.21

4 $\vec{AB} = \mathbf{b} - \mathbf{a}$, so $\vec{EB} = \frac{1}{2}(\mathbf{b} - \mathbf{a})$; and
$\vec{BC} = \mathbf{c} - \mathbf{b}$; so $\vec{BF} = \frac{1}{2}(\mathbf{c} - \mathbf{b})$. Therefore
$\vec{EF} = \vec{EB} + \vec{BF} = \frac{1}{2}(\mathbf{b} - \mathbf{a}) + \frac{1}{2}(\mathbf{c} - \mathbf{b}) = \frac{1}{2}(\mathbf{c} - \mathbf{a})$. Also,
$\vec{AD} = \mathbf{d} - \mathbf{a}$, so $\vec{HD} = \frac{1}{2}(\mathbf{d} - \mathbf{a})$; and
$\vec{DC} = \mathbf{c} - \mathbf{d}$, so $\vec{DG} = \frac{1}{2}(\mathbf{c} - \mathbf{d})$. Therefore
$\vec{HG} = \vec{HD} + \vec{DG} = \frac{1}{2}(\mathbf{d} - \mathbf{a}) + \frac{1}{2}(\mathbf{c} - \mathbf{d}) = \frac{1}{2}(\mathbf{c} - \mathbf{a})$.
Hence $\vec{EF} = \vec{HG}$.

Exercises 1.2

1 $\alpha(1, 1, 1) + \beta(1, 2, 3) + \gamma(1, -1, -3) = (0, 0, 0)$
$\Rightarrow \alpha + \beta + \gamma = 0$, $\alpha + 2\beta - \gamma = 0$, $\alpha + 3\beta - 3\gamma = 0$
$\Rightarrow \alpha = 3$, $\beta = -2$, $\gamma = -1$. Hence
$$3(1, 1, 1) - 2(1, 2, 3) - (1, -1, -3) = (0, 0, 0).$$

2 $(1, 1, 1) = (2/3)(1, 2, 3) + (1/3)(1, -1, -3)$,
$(1, 2, 3) = (3/2)(1, 1, 1) - (1/2)(1, -1, -3)$,
$(1, -1, -3) = 3(1, 1, 1) - 2(1, 2, 3)$.

3 Linear independence is straightforward. Also,
$(-1, 2, 3) = -(1, -2, 0) + 3(0, 1, 4) + 3(0, -1, -3)$.

4 $3(-1, 1, 2) + 0(1, 1, 0) + (3, -3, -6) = (0, 0, 0)$.
Clearly $3(-1, 1, 2) = -1(3, -3, -6)$.
However, $(1, 1, 0) = \alpha(-1, 1, 2) + \beta(3, -3, -6) = (\alpha - 3\beta)(-1, 1, 2)$ implies that $\alpha - 3\beta = -1 = 1 = 0$, which is impossible.

5 For example, $(-1, 1, 2)$, $(1, 1, 0)$ are linearly independent. A possible third vector in this case is, for instance, $(0, 0, 1)$.

6 Simply check that they are linearly independent.

Exercises 1.3

1 (a) 3; (b) $3/\sqrt{2}\sqrt{5} = 3/\sqrt{10}$.

2 (a) 3; (b) $3/\sqrt{3}\sqrt{5} = 3/\sqrt{15} = \sqrt{3/5}$.

3 $|(1, 1)| = \sqrt{2}$, $|(1, 2)| = \sqrt{5}$, $|(1, 1, 1)| = \sqrt{3}$, $|(1, 2, 0)| = \sqrt{5}$.

4 Let $\mathbf{a} = (x, y, z)$ be orthogonal to the two vectors. Then
$0 = (x, y, z) \cdot (1, 2, 4) = x + 2y + 4z$, $0 = (x, y, z) \cdot (2, 1, -1) = 2x + y - z$.

Solving gives $y = -3z$, $x = 2z$. Hence $\mathbf{a} = z(2, -3, 1)$ where z is any non-zero real number.

5 By the method of question 4, $\mathbf{a} = y(2, 1, -5)$, where $0 \neq y \in \mathbb{R}$, is orthogonal to the given vectors. In particular, taking $y = 1$ gives $(2, 1, -5)$. Checking for linear independence is straightforward.

6 (Fig. 1.22) Let $\mathbf{a}, \mathbf{b}, \mathbf{c}$ ($= \mathbf{a} + \mathbf{b}$) be as shown. Then
$c^2 = \mathbf{c} \cdot \mathbf{c} = (\mathbf{a} + \mathbf{b}) \cdot (\mathbf{a} + \mathbf{b})$
$= \mathbf{a} \cdot \mathbf{a} + \mathbf{a} \cdot \mathbf{b} + \mathbf{b} \cdot \mathbf{a} + \mathbf{b} \cdot \mathbf{b}$
$= a^2 + 2ab \cos(\pi - \gamma) + b^2$
$= a^2 - 2ab \cos \gamma + b^2$.

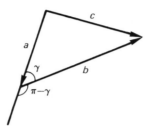

Fig. 1.22

7 (Fig. 1.23) $\mathbf{x} - \mathbf{y}$ and $\mathbf{x} + \mathbf{y}$ are orthogonal
$\Leftrightarrow 0 = (\mathbf{x} - \mathbf{y}) \cdot (\mathbf{x} + \mathbf{y}) = \mathbf{x} \cdot \mathbf{x} + \mathbf{x} \cdot \mathbf{y} - \mathbf{y} \cdot \mathbf{x} - \mathbf{y} \cdot \mathbf{y} = |\mathbf{x}|^2 - |\mathbf{y}|^2$
$\Leftrightarrow |\mathbf{x}| = |\mathbf{y}|$ (since $|\mathbf{x}|, |\mathbf{y}| > 0$).
For the last part note that $\vec{OZ} = \mathbf{x} + \mathbf{y}$, $\vec{YX} = \mathbf{x} - \mathbf{y}$.

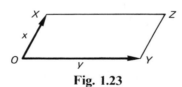

Fig. 1.23

8 Straightforward (if tedious!).

Exercises 1.4

1 (a) $(0, 0, 3)$; (b) $|\mathbf{v}| = \sqrt{2}, |\mathbf{w}| = \sqrt{29}, |\mathbf{v} \times \mathbf{w}| = 3$;
   (c) $\sin \theta = |\mathbf{v} \times \mathbf{w}|/|\mathbf{v}||\mathbf{w}| = 3/\sqrt{2}\sqrt{29} = 3/\sqrt{58}$.

2 (a) $(2, 1, 3)$; (b) $|\mathbf{v}| = \sqrt{3}, |\mathbf{w}| = \sqrt{5}, |\mathbf{v} \times \mathbf{w}| = \sqrt{14}$;
   (c) $\sin \theta = \sqrt{14}/\sqrt{3}\sqrt{5} = \sqrt{14/15}$.

3 We have $(0, -1, 2) \times (3, 1, 1) = (-3, 6, 3) = 3(-1, 2, 1)$.
   Now $|(-1, 2, 1)| = \sqrt{6}$, so the required unit vector is
   $\pm(-1/\sqrt{6}, 2/\sqrt{6}, 1/\sqrt{6})$.

4 (Fig. 1.24) Let **a, b, c** be as shown. Then $\mathbf{a}+\mathbf{b}+\mathbf{c}=\mathbf{0}$,

so
$$\mathbf{0} = \mathbf{a} \times \mathbf{0} = \mathbf{a} \times (\mathbf{a}+\mathbf{b}+\mathbf{c})$$
$$= \mathbf{a} \times \mathbf{a} + \mathbf{a} \times \mathbf{b} + \mathbf{a} \times \mathbf{c} = \mathbf{a} \times \mathbf{b} + \mathbf{a} \times \mathbf{c}.$$

Hence $\mathbf{b} \times \mathbf{a} = \mathbf{a} \times \mathbf{c}$. Equating lengths gives $ab \sin \gamma = ac \sin \beta$. Expanding $\mathbf{b} \times (\mathbf{a}+\mathbf{b}+\mathbf{c})$ completes the demonstration.

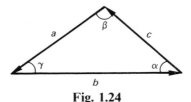

Fig. 1.24

5 Straightforward (if boring!).

6 Simply (!) write out the left-hand side in terms of coordinates, and all will be revealed! Alternatively (but less obviously) notice that it suffices to check the identity for a basis, and that most of the combinations of **i, j, k** can be checked mentally.

Exercises 1.5

1 The vector equation is $(x, y, z) = (1, 2, 3) + \lambda(1, 1, 0)$. Thus $x = 1 + \lambda$, $y = 2 + \lambda$, $z = 3$, whence the equation(s) is (are) $x = y - 1$, $z = 3$.

2 The vector equation is $(x, y, z) = (1 - \lambda)(1, 1, 1) + \lambda(1, 1, 2)$. Thus $x = 1 - \lambda + \lambda = 1$, $y = 1 - \lambda + \lambda = 1$, $z = 1 - \lambda + 2\lambda = 1 + \lambda$, whence the equation(s) is (are) $x = 1, y = 1$. (Note that the $z = 1 + \lambda$ adds nothing since λ can take any real value, and hence z can also take any real value; this is the line parallel to the z-axis through the point $(1, 1, 1)$.)

3 The vector equation is $(x, y, z) = (1 - \lambda - \mu)(1, 0, 0) + \lambda(0, 1, 0) + \mu(0, 0, 1)$. Thus $x = 1 - \lambda - \mu$, $y = \lambda$, $z = \mu$, whence the equation is $x + y + z = 1$.

4 The equation is of the form $q = (x, y, z) \cdot (1, 1, 0) = x - y$. Since $(1, 2, 1)$ lies on the plane, $q = 1 - 2 = -1$. The equation is therefore $x - y = -1$. Now $|(1, -1, 0)| = \sqrt{2}$, so the equation in the form $\mathbf{r} \cdot \hat{\mathbf{n}} = p$ is $(x, y, z) \cdot (1/\sqrt{2}, -1/\sqrt{2}, 0) = -1/\sqrt{2}$. Thus, the distance from the origin is $1/\sqrt{2}$.

5 The vector $(1, 2, 3) \times (-4, 2, 2) = (-2, -14, 10)$ is perpendicular to the plane, so $\pm(1/5\sqrt{3}, 7/5\sqrt{3}, -1/\sqrt{3})$ is the required unit vector. The required distance is $(2, 3, 4) \cdot (1/5\sqrt{3}, 7/5\sqrt{3}, -1/\sqrt{3}) = \sqrt{3}/5$, since the distance from the origin is 0.

6 (a) AB has equation $(x, y, z) = (1 - \lambda)(0, 0, 3) + \lambda(1, 1, 5) = (\lambda, \lambda, 3 + 2\lambda)$;
CD has equation $(x, y, z) = (1 - \mu)(0, 3, 0) + \mu(-1, 5, 1)$
$= (-\mu, 3 + 2\mu, \mu)$.

(b) $\vec{AB} = (1, 1, 2)$, $\vec{CD} = (-1, 2, 1)$. A unit vector orthogonal to both of them is $\pm(1/\sqrt{3}, 1/\sqrt{3}, -1/\sqrt{3})$.

(c) Let P be the point $(\lambda, \lambda, 3 + 2\lambda)$ (which lies on AB), and let Q be the point $(-\mu, 3 + 2\mu, \mu)$ (which lies on CD).
Then $\vec{QP} = (\lambda + \mu, \lambda - 3 - 2\mu, 3 + 2\lambda - \mu)$. This must be parallel to the unit vector of (b), so we want $\vec{QP} = z(1, 1, -1)$.
Hence $\lambda + \mu = \lambda - 3 - 2\mu = \mu - 3 - 2\lambda$. Solving gives $\mu = -1, \lambda = -1$.
Thus P is $(-1, -1, 1)$, Q is $(1, 1, -1)$. Finally,
$|\vec{PQ}| = \vec{AC}\cdot\hat{\mathbf{n}} = \vec{BD}\cdot\hat{\mathbf{n}} = \sqrt{12}$.

7 (Fig. 1.25)

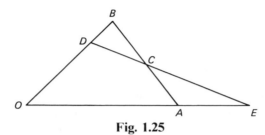

Fig. 1.25

(a) $\mathbf{c} = \vec{OC} = \tfrac{1}{2}(\mathbf{a} + \mathbf{b})$, $\mathbf{d} = \vec{OD} = \tfrac{3}{4}\mathbf{b}$, so CD has equation
$\mathbf{r} = (1 - \lambda)\mathbf{c} + \lambda\mathbf{d} = \tfrac{1}{2}(1 - \lambda)(\mathbf{a} + \mathbf{b}) + \tfrac{3}{4}\lambda\mathbf{b} = \tfrac{1}{2}(1 - \lambda)\mathbf{a} + \tfrac{1}{4}(2 + \lambda)\mathbf{b}$.

(b) Let $\mathbf{e} = \vec{OE}$. Then, since E lies on OA and on CD,
$\mathbf{e} = \mu\mathbf{a} = \tfrac{1}{2}(1 - \lambda)\mathbf{a} + \tfrac{1}{4}(2 + \lambda)\mathbf{b}$. Hence $(\mu - \tfrac{1}{2}(1 - \lambda))\mathbf{a} + \tfrac{1}{4}(2 + \lambda)\mathbf{b} = \mathbf{0}$.
Since $\mathbf{a}$ and $\mathbf{b}$ do not lie along the same straight line they are linearly independent, and so $\mu - \tfrac{1}{2}(1 - \lambda) = 0, \tfrac{1}{4}(2 + \lambda) = 0$. Thus $\lambda = -2, \mu = \tfrac{3}{2}$, and $\vec{OE} = \tfrac{3}{2}\mathbf{a}$.

(c) $\vec{CD}$ and $\vec{OD}$ are orthogonal $\Rightarrow \vec{CD}\cdot\vec{OD} = 0$
$\Rightarrow 0 = (\mathbf{d} - \mathbf{c})\cdot\mathbf{d} = (\tfrac{1}{4}\mathbf{b} - \tfrac{1}{2}\mathbf{a})\cdot\tfrac{3}{4}\mathbf{b} = \tfrac{3}{16}(\mathbf{b}\cdot\mathbf{b} - 2\mathbf{a}\cdot\mathbf{b})$
$= \tfrac{3}{16}(|\mathbf{b}|^2 - 2|\mathbf{a}||\mathbf{b}|\cos\widehat{AOB}) \Rightarrow \cos\widehat{AOB} = |\mathbf{b}|/2|\mathbf{a}|$.

8 (Fig. 1.26) Let $\mathbf{a}, \mathbf{b}, \mathbf{c}, \mathbf{p}, \mathbf{r}, \mathbf{s}$ be the position vectors of A, B, C, P, R, S respectively with respect to O. Then any point on AS is of the form $(1 - \lambda)\mathbf{a} + \lambda\mathbf{s} = (1 - \lambda)\mathbf{a} + \tfrac{1}{2}\lambda\mathbf{c}$; any point on PB has the form $(1 - \mu)\mathbf{p} + \mu\mathbf{b} = \tfrac{1}{2}(1 - \mu)\mathbf{a} + \mu(\mathbf{a} + \mathbf{c})$. Hence
$\mathbf{r} = (1 - \lambda)\mathbf{a} + \tfrac{1}{2}\lambda\mathbf{c} = \tfrac{1}{2}(1 - \mu)\mathbf{a} + \mu(\mathbf{a} + \mathbf{c})$. As in question 7 we can equate coefficients of $\mathbf{a}, \mathbf{c}$ on each side to give $(1 - \lambda) = \tfrac{1}{2}(1 + \mu), \tfrac{1}{2}\lambda = \mu$. Therefore $\mu = \tfrac{1}{5}, \lambda = \tfrac{2}{5}$ and $\mathbf{r} = \tfrac{3}{5}\mathbf{a} + \tfrac{1}{5}\mathbf{c} = \tfrac{2}{5}\mathbf{a} + \tfrac{1}{5}(\mathbf{a} + \mathbf{c}) = \tfrac{2}{5}\mathbf{a} + \tfrac{1}{5}\mathbf{b} = \tfrac{4}{5}\mathbf{p} + \tfrac{1}{5}\mathbf{b}$. It follows from exercises 1.1.1 (e) that $PR:RB = 1:4$.

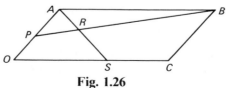

Fig. 1.26

9 (Fig. 1.27)

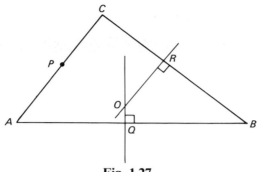

Fig. 1.27

Let OR, OQ be the perpendicular bisectors of BC, AB, respectively, and take the intersection of OR, OQ as origin. Then $\overrightarrow{OQ} = \frac{1}{2}(\mathbf{a} + \mathbf{b})$, $\overrightarrow{OR} = \frac{1}{2}(\mathbf{b} + \mathbf{c})$, $\overrightarrow{AB} = \mathbf{b} - \mathbf{a}$, $\overrightarrow{BC} = \mathbf{c} - \mathbf{b}$. Let P be the mid-point of AC. Because OQ is a perpendicular bisector,
$0 = \frac{1}{2}(\mathbf{a} + \mathbf{b}) \cdot (\mathbf{b} - \mathbf{a}) = \frac{1}{2}\mathbf{a} \cdot \mathbf{b} + \frac{1}{2}\mathbf{b} \cdot \mathbf{b} - \frac{1}{2}\mathbf{a} \cdot \mathbf{a} - \frac{1}{2}\mathbf{b} \cdot \mathbf{a} \Rightarrow |\mathbf{a}|^2 = |\mathbf{b}|^2$.
Similarly, because OR is a perpendicular bisector, $\frac{1}{2}(\mathbf{b} + \mathbf{c}) \cdot (\mathbf{c} - \mathbf{b}) = 0$ implies that $|\mathbf{b}|^2 = |\mathbf{c}|^2$. Thus, $|\mathbf{a}|^2 = |\mathbf{b}|^2 = |\mathbf{c}|^2$. We need to show that OP is perpendicular to AC. Now $\overrightarrow{OP} = \frac{1}{2}(\mathbf{a} + \mathbf{c})$, $\overrightarrow{AC} = \mathbf{c} - \mathbf{a}$, and so $\frac{1}{2}(\mathbf{a} + \mathbf{c}) \cdot (\mathbf{c} - \mathbf{a}) = \frac{1}{2}(|\mathbf{c}|^2 - |\mathbf{a}|^2) = 0$, as required.

10 Take as origin the circumcentre of the triangle. Then, as in question 9 above, $|\mathbf{a}|^2 = |\mathbf{b}|^2 = |\mathbf{c}|^2$. It suffices to show that the point $(\mathbf{a} + \mathbf{b} + \mathbf{c})$ lies on each of the altitudes. Consider the line from C passing through the point $\mathbf{a} + \mathbf{b} + \mathbf{c}$. This line is in the direction of $\mathbf{a} + \mathbf{b} + \mathbf{c} - \mathbf{c} = \mathbf{a} + \mathbf{b}$. But the vector AB is $\mathbf{b} - \mathbf{a}$, and $(\mathbf{a} + \mathbf{b}) \cdot (\mathbf{b} - \mathbf{a}) = |\mathbf{b}|^2 - |\mathbf{a}|^2 = 0$. So the line from C through the point $\mathbf{a} + \mathbf{b} + \mathbf{c}$ is indeed an altitude. Similarly the lines from A and B through $\mathbf{a} + \mathbf{b} + \mathbf{c}$ are altitudes.

11 We saw in question 10 above that, if the origin is at the circumcentre, then the orthocentre is at $\mathbf{a} + \mathbf{b} + \mathbf{c}$. But we saw in example 1.5.2 that the centroid is at $\frac{1}{3}(\mathbf{a} + \mathbf{b} + \mathbf{c})$, regardless of where we choose the origin to be. The result follows from these two observations.

12 Any point on the line CF is of the form $(1 - \lambda)\mathbf{c} + \lambda(\frac{1}{2}(\mathbf{a} + \mathbf{b}))$. Putting $\lambda = \frac{2}{3}$ shows that G lies on CF.

2 MATRICES

2.1 INTRODUCTION

In this chapter we introduce another sort of quantity which can be manipulated formally in much the same way that vectors and polynomials can.

An $m \times n$ real (complex) matrix A is an array of real (complex) numbers a_{ij} ($1 \leq i \leq m$, $1 \leq j \leq n$) arranged in m rows and n columns and enclosed by brackets, as follows:

$$\begin{bmatrix} a_{11} & a_{12} & \cdots & a_{1n} \\ a_{21} & a_{22} & \cdots & a_{2n} \\ \vdots & & & \vdots \\ a_{m1} & a_{m2} & \cdots & a_{mn} \end{bmatrix}. \qquad (\dagger)$$

Notes:

(a) If A is $m \times n$, then m is the number of rows in the array.

(b) If A is $m \times n$, then n is the number of columns in the array.

(c) The *size* (or *order*) of the matrix is $m \times n$.

(d) a_{ij} appears in the ith row and the jth column.

(e) The numbers a_{ij} are called the *elements* of the matrix.

(f) The notation ($\dagger$) is sometimes abbreviated to $[a_{ij}]$, or to $[a_{ij}]_{m \times n}$ if we wish to specify the size of the array.

Examples 2.1.1

1. $A = \begin{bmatrix} 1 & 2 & 3 \\ 4 & 5 & 6 \end{bmatrix}$ is a 2×3 matrix in which $a_{11} = 1, a_{12} = 2, a_{13} = 3$, $a_{21} = 4, a_{22} = 5, a_{23} = 6$.

2. $A = \begin{bmatrix} 1 & 2 \\ 3 & 4 \\ 5 & 6 \end{bmatrix}$ is a 3×2 matrix in which $a_{11} = 1, a_{12} = 2, a_{21} = 3$, $a_{22} = 4, a_{31} = 5, a_{33} = 6$.

3. $A = \begin{bmatrix} 0 & 6i \\ i-5 & -9 \end{bmatrix}$ is a 2×2 matrix in which $a_{11} = 0, a_{12} = 6i, a_{21} = i - 5$, $a_{22} = -9$.

Matrices arise in a number of contexts in mathematics. In particular, we will examine their use later in solving systems of simultaneous linear equations. Given a homogeneous system such as

$$3x + 2y + 5z = 0$$
$$-x + 7y - 3z = 0$$
$$2x - y - z = 0$$

we can associate with it a matrix whose elements are the coefficients of x, y, z in the equations, namely

$$\begin{bmatrix} 3 & 2 & 5 \\ -1 & 7 & -3 \\ 2 & -1 & -1 \end{bmatrix}$$

This matrix of coefficients contains all of the essential information about the equations, and we will learn how to solve the system, and corresponding inhomogeneous systems, by operating on the matrix. First we need to become proficient in matrix algebra.

Two matrices $A = [a_{ij}]_{m \times n}$, $B = [b_{ij}]_{r \times s}$ are *equal* if $m = r, n = s$ and $a_{ij} = b_{ij}$ for $1 \leq i \leq m(=r), 1 \leq j \leq n(=s)$; that is, if they have the same number of rows, the same number of columns, and corresponding elements are equal.

Examples 2.1.2

1. $\begin{bmatrix} 1 & 2 \\ 3 & 4 \end{bmatrix} \neq \begin{bmatrix} 1 & 3 \\ 2 & 4 \end{bmatrix}$ since $2 \neq 3$.

2. $\begin{bmatrix} 1 & 2 & 0 \\ 3 & 4 & 0 \end{bmatrix} \neq \begin{bmatrix} 1 & 2 \\ 3 & 4 \end{bmatrix}$ as a 2×3 matrix cannot be equal to a 2×2 matrix.

An $m \times n$ matrix A is *square* if $m = n$; that is, if A has the same number of rows and columns. In a square matrix $A = [a_{ij}]_{n \times n}$ the elements $a_{11}, a_{22}, \ldots, a_{nn}$ are called the elements of the *main* (or *leading*) *diagonal*.

A $1 \times n$ matrix $[a_{11}\ a_{12} \ldots a_{1n}]$ is often referred to as a *row vector*, because of the resemblance to a vector specified in coordinate form. Similarly, an $m \times 1$ matrix $\begin{bmatrix} a_{11} \\ a_{21} \\ \cdot \\ a_{m1} \end{bmatrix}$ may be called a *column vector*.

EXERCISES 2.1

1 Write down the sizes of the following matrices:

(a) $\begin{bmatrix} i & -i \\ 1 & 2 \\ 1+i & 4 \\ \frac{1}{2} & -1 \end{bmatrix}$
(b) $[1\ \ -1\ \ 1\ \ -1\ \ 1\ \ -1]$

(c) $\begin{bmatrix} 1 & 3 & 5 \\ 2 & -1 & -4 \\ 6 & 1 & 3 \end{bmatrix}$
(d) $\begin{bmatrix} 1 & -1 \\ -1 & 1 \end{bmatrix}$.

2 Which of the matrices in question 1 above are square? For each that is square, write down the elements of its main diagonal.

3 Write down the column vector whose elements form the second column of the matrix in question 1 (c) above.

4 Write down the row vector whose elements form the third row of the matrix in question 1 (c) above.

2.2 ADDITION AND SCALAR MULTIPLICATION OF MATRICES

If $A = [a_{ij}]_{m \times n}$, $B = [b_{ij}]_{m \times n}$ are two $m \times n$ matrices, their *sum*, $A + B$, is defined to be the matrix $[a_{ij} + b_{ij}]_{m \times n}$. Notice that this sum is defined only when the two matrices have the same number of rows and the same number of columns. The sum is formed by adding corresponding elements, and produces a matrix of the same size as A and B. Similarly, the *difference*, $A - B$, is $[a_{ij} - b_{ij}]_{m \times n}$.

Examples 2.2.1

1. $\begin{bmatrix} 1 & 2 \\ 3 & 4 \end{bmatrix} + \begin{bmatrix} 3 & 1 \\ -1 & 5 \end{bmatrix} = \begin{bmatrix} 1+3 & 2+1 \\ 3+(-1) & 4+5 \end{bmatrix} = \begin{bmatrix} 4 & 3 \\ 2 & 9 \end{bmatrix}$.

2. $\begin{bmatrix} 1 & -1 \\ 2 & 6 \\ -3 & 1 \end{bmatrix} + \begin{bmatrix} 2 & 1 \\ 1 & -2 \\ 3 & 0 \end{bmatrix} = \begin{bmatrix} 1+2 & (-1)+1 \\ 2+1 & 6+(-2) \\ (-3)+3 & 1+0 \end{bmatrix} = \begin{bmatrix} 3 & 0 \\ 3 & 4 \\ 0 & 1 \end{bmatrix}.$

3. $\begin{bmatrix} 1 & 2 \\ 1 & -1 \end{bmatrix} + \begin{bmatrix} 3 & 1 & 0 \\ 4 & 1 & 0 \end{bmatrix}$ is not defined.

4. $\begin{bmatrix} 1 & 2 \\ 3 & 4 \end{bmatrix} - \begin{bmatrix} 1 & 0 \\ -1 & 3 \end{bmatrix} = \begin{bmatrix} 1-1 & 2-0 \\ 3-(-1) & 4-3 \end{bmatrix} = \begin{bmatrix} 0 & 2 \\ 4 & 1 \end{bmatrix}.$

5. $\begin{bmatrix} 1 \\ 2 \end{bmatrix} - \begin{bmatrix} 1 & -1 \end{bmatrix}$ is not defined.

If α is a *scalar* (a real or complex number) and $A = [a_{ij}]_{m \times n}$ is an $m \times n$ matrix, then αA is defined to be the $m \times n$ matrix $[\alpha a_{ij}]_{m \times n}$. This scalar multiplication by α, therefore, is performed simply by multiplying each element of A by α. We write $-A$ for $(-1)A$, and call $-A$ the *negative* (or *additive inverse*) of A.

Examples 2.2.2

1. $3 \begin{bmatrix} 1 & -1 \\ 0 & 2 \end{bmatrix} = \begin{bmatrix} 3 \times 1 & 3 \times (-1) \\ 3 \times 0 & 3 \times 2 \end{bmatrix} = \begin{bmatrix} 3 & -3 \\ 0 & 6 \end{bmatrix}.$

2. $-1 \begin{bmatrix} 1 & 1 & -1 \\ 2 & 0 & 3 \end{bmatrix} = \begin{bmatrix} -1 \times 1 & -1 \times 1 & -1 \times -1 \\ -1 \times 2 & -1 \times 0 & -1 \times 3 \end{bmatrix} = \begin{bmatrix} -1 & -1 & 1 \\ -2 & 0 & -3 \end{bmatrix}.$

3. $\frac{1}{2} \begin{bmatrix} 4 \\ 2 \\ 6 \end{bmatrix} = \begin{bmatrix} 2 \\ 1 \\ 3 \end{bmatrix}.$

A matrix all of whose elements are zero is called a *zero matrix*. We denote such a matrix by $O_{m \times n}$, or simply by O if there can be no confusion about its size. Then addition and scalar multiplication of matrices satisfies the following properties:

Let A, B, C be any real or complex matrices, and let α, β, γ be any real or complex numbers.

V1 $(A + B) + C = A + (B + C)$.
V2 $A + O = O + A = A$.
V3 $A + (-A) = (-A) + A = O$.
V4 $A + B = B + A$.
V5 $\alpha(A + B) = \alpha A + \alpha B$.
V6 $(\alpha + \beta)A = \alpha A + \beta A$.
V7 $(\alpha\beta)A = \alpha(\beta A)$.
V8 $1A = A$.
V9 $0A = O$.

In V2, V3, and V9 it is implicit that O has the same size as A. All of the properties are straightforward to check. For example, consider V1.

Let $\qquad A = [a_{ij}]_{m \times n}, \ B = [b_{ij}]_{m \times n}, \ C = [c_{ij}]_{m \times n}.$

Then

$$(A + B) + C = [(a_{ij} + b_{ij}) + c_{ij}]_{m \times n}, \qquad A + (B + C) = [a_{ij} + (b_{ij} + c_{ij})]_{m \times n}$$

and these are equal because of the associative property of addition of real and complex numbers (that is, $(a_{ij} + b_{ij}) + c_{ij} = a_{ij} + (b_{ij} + c_{ij})$). The others are left as an easy exercise for the reader.

It is hoped that the reader will have a strong sense of having seen V1 to V9 before. They are, of course, the very same properties that were found for addition and scalar multiplication of vectors in Chapter 1. The full significance of this fact will be exposed in a later chapter; for the time being we simply record our observation that the same properties appear in these two apparently different contexts.

EXERCISES 2.2

For each of the following, decide whether it is defined and, if it is, compute the result:

1 $\begin{bmatrix} 1 \\ 2 \end{bmatrix} + \begin{bmatrix} 2 \\ -1 \end{bmatrix}$

2 $[1 \quad -3] + [4 \quad -1]$

3 $\begin{bmatrix} 1 \\ 2 \end{bmatrix} + [3 \quad 2]$

4 $\begin{bmatrix} 1+i & 3+2i \\ 2-i & i \end{bmatrix} - \begin{bmatrix} 4-i & -i \\ 2 & 0 \end{bmatrix}$

5 $\begin{bmatrix} 1 & -1 \\ 3 & 4 \\ 2 & 1 \end{bmatrix} - \begin{bmatrix} 2 & 1 \\ 3 & 2 \end{bmatrix}$

6 $\begin{bmatrix} 1 & 1 \\ 5 & -2 \\ 1 & -1 \end{bmatrix} - \begin{bmatrix} 2 & 1 \\ 6 & 3 \\ 0 & 0 \end{bmatrix}$

7 $\quad 3i \begin{bmatrix} i & -1 \\ 4-i & \frac{1}{2} \end{bmatrix}$
  $\quad\quad$ **8** $\quad \frac{1}{2} \begin{bmatrix} 2 & -2 \\ 4 & 6 \end{bmatrix}$

9 $\quad 3 \begin{bmatrix} 1 & 2 & 1 & 4 \\ 8 & 4 & 9 & 1 \\ 1 & 5 & 0 & 0 \\ 0 & 4 & 2 & 7 \end{bmatrix} - 2 \begin{bmatrix} 1 & -1 & 4 & 6 \\ 0 & -1 & 5 & 3 \\ 4 & -3 & 3 & 0 \\ 6 & 0 & 4 & 1 \end{bmatrix}$

10 $\quad 4 \begin{bmatrix} 1 & -1 \\ -1 & 1 \end{bmatrix} + 2 \begin{bmatrix} 1 & 1 \\ -1 & 2 \\ 0 & 0 \end{bmatrix}$

11 $\quad i[i \quad 1-i] - (1+i)[1 \quad -i]$
$\quad\quad$ **12** $\quad 2 \begin{bmatrix} 0 \\ -1 \end{bmatrix} - 3 \begin{bmatrix} -1 \\ -2 \end{bmatrix}$

2.3 MATRIX MULTIPLICATION

Our next task is to define a means of multiplying two matrices together. An obvious definition which we might try would be to consider only products of matrices of the same size and simply to multiply together corresponding elements; that is, if $A = [a_{ij}]_{m \times n}$, $B = [b_{ij}]_{m \times n}$, then to define AB to be $[a_{ij}b_{ij}]_{m \times n}$. However, this will not suffice for the applications we have in mind; there is a method which at first sight appears to be more cumbersome, but which ultimately rewards us for our extra efforts. The idea is related to that of the scalar product of two vectors. Let us consider a little more carefully one of the applications which has been promised: namely, to systems of linear equations. The system

$$a_{11}x_1 + \ldots + a_{1n}x_n = c_1$$
$$\ldots\ldots\ldots\ldots\ldots\ldots\ldots\ldots \quad (\dagger)$$
$$a_{m1}x_1 + \ldots + a_{mn}x_n = c_m$$

can be written as
$$\mathbf{a}_1 \cdot \mathbf{x} = c_1$$
$$\ldots\ldots\ldots$$
$$\mathbf{a}_m \cdot \mathbf{x} = c_m$$

where $\mathbf{a}_i = (a_{i1}, \ldots, a_{in})$ for $1 \leq i \leq m$, and $\mathbf{x} = (x_1, \ldots, x_n)$.

We could write it more simply still as

$$AX = C$$

where $\quad A = \begin{bmatrix} a_{11} & \ldots & a_{1n} \\ \ldots\ldots\ldots\ldots \\ a_{m1} & \ldots & a_{mn} \end{bmatrix}, \quad X = \begin{bmatrix} x_1 \\ \vdots \\ x_n \end{bmatrix}, \quad C = \begin{bmatrix} c_1 \\ \vdots \\ c_m \end{bmatrix},$

if we were to define AX to be equal to $\begin{bmatrix} \mathbf{a}_1 \cdot \mathbf{x} \\ \vdots \\ \mathbf{a}_m \cdot \mathbf{x} \end{bmatrix}$. This gives us a definition for the product of an $m \times n$ matrix and an $n \times 1$ matrix.

Now suppose we were to make the following linear substitution:

$$x_1 = b_{11}y_1 + \ldots + b_{1p}y_p$$
$$\ldots\ldots\ldots\ldots\ldots\ldots\ldots\ldots\ldots$$
$$x_n = b_{n1}y_1 + \ldots + b_{np}y_p$$

This could be written in matrix form as

$$X = BY$$

where $\quad B = \begin{bmatrix} b_{11} & \ldots & b_{1p} \\ \ldots & & \ldots \\ b_{n1} & \ldots & b_{np} \end{bmatrix}$ and $\quad Y = \begin{bmatrix} y_1 \\ \vdots \\ y_p \end{bmatrix}$.

We would wish the result to be given by $C = AX = A(BY) = (AB)Y$. Now, if we make this substitution into the system (†) we obtain

$$a_{11}(b_{11}y_1 + \ldots + b_{1p}y_p) + \ldots + a_{in}(b_{n1}y_1 + \ldots + b_{np}y_p) = 0$$
$$\ldots\ldots\ldots\ldots\ldots\ldots\ldots\ldots\ldots\ldots\ldots\ldots\ldots\ldots\ldots\ldots\ldots$$
$$a_{m1}(b_{11}y_1 + \ldots + b_{1p}y_p) + \ldots + a_{mn}(b_{n1}y_1 + \ldots + b_{np}y_p) = 0.$$

Hence, collecting together the coefficients of y_i,

$$(a_{11}b_{11} + \ldots + a_{1n}b_{n1})y_1 + \ldots + (a_{11}b_{1p} + \ldots + a_{1n}b_{np})y_p = 0$$
$$\ldots\ldots\ldots\ldots\ldots\ldots\ldots\ldots\ldots\ldots\ldots\ldots\ldots\ldots\ldots\ldots\ldots$$
$$(a_{m1}b_{11} + \ldots + a_{mn}b_{n1})y_1 + \ldots + (a_{m1}b_{1p} + \ldots + a_{mn}b_{np})y_p = 0.$$

We therefore define AB to be given by

$$AB = \begin{bmatrix} a_{11}b_{11} + \ldots + a_{1n}b_{n1} & \ldots & a_{11}b_{1p} + \ldots + a_{1n}b_{np} \\ \ldots & & \ldots \\ a_{m1}b_{11} + \ldots + a_{mn}b_{n1} & \ldots & a_{m1}b_{1p} + \ldots + a_{mn}b_{np} \end{bmatrix}$$

$$= \begin{bmatrix} \mathbf{a}_1 \cdot \mathbf{b}_1 & \ldots & \mathbf{a}_1 \cdot \mathbf{b}_p \\ \ldots & & \ldots \\ \mathbf{a}_m \cdot \mathbf{b}_1 & \ldots & \mathbf{a}_m \cdot \mathbf{b}_p \end{bmatrix},$$

where $\mathbf{b}_i = (b_{1i}, \ldots, b_{ni})$ for $1 \leq i \leq p$. Here $\mathbf{a}_1, \ldots, \mathbf{a}_m$ come from the rows of A, and $\mathbf{b}_1, \ldots, \mathbf{b}_p$ come from the columns of B.

MATRICES

Note:
$$\underset{m \times n}{A} \underset{n \times p}{B} = \underset{m \times p}{C}$$

This product is defined only when *the number of columns of* A *is equal to the number of rows of* B.

If $A = [a_{ij}]_{m \times n}$, $B = [b_{ij}]_{r \times p}$, then AB is defined if and only if $n = r$, and in this case

$$AB = \left[\sum_{k=1}^{n} a_{ik}b_{kj} \right]_{m \times p}.$$

Clearly, such a complicated-looking definition takes a little while to absorb, but, with practice, matrix multiplications become second nature.

Examples 2.3

1. $\begin{bmatrix} 1 & 2 \\ 3 & 4 \end{bmatrix} \begin{bmatrix} 5 & 6 \\ 7 & 8 \end{bmatrix} = \begin{bmatrix} 1 \times 5 + 2 \times 7 & 1 \times 6 + 2 \times 8 \\ 3 \times 5 + 4 \times 7 & 3 \times 6 + 4 \times 8 \end{bmatrix} = \begin{bmatrix} 19 & 22 \\ 43 & 50 \end{bmatrix}.$
(2 × 2) (2 × 2)   (2 × 2)

2. $\begin{bmatrix} 1 & -1 & 0 \\ 2 & 1 & 3 \\ 0 & 0 & 1 \end{bmatrix} \begin{bmatrix} 3 & 1 & 1 \\ -1 & -1 & 0 \\ 4 & 1 & 3 \end{bmatrix}$
  (3 × 3)   (3 × 3)

$= \begin{bmatrix} 1 \times 3 + (-1)(-1) + 0 \times 4 & 1 \times 1 + (-1)(-1) + 0 \times 1 & 1 \times 1 + (-1)0 + 0 \times 3 \\ 2 \times 3 + 1(-1) + 3 \times 4 & 2 \times 1 + 1(-1) + 3 \times 1 & 2 \times 1 + 1 \times 0 + 3 \times 3 \\ 0 \times 3 + 0(-1) + 1 \times 4 & 0 \times 1 + 0(-1) + 1 \times 1 & 0 \times 1 + 0 \times 0 + 1 \times 3 \end{bmatrix}$

$= \begin{bmatrix} 4 & 2 & 1 \\ 17 & 4 & 11 \\ 4 & 1 & 3 \end{bmatrix}.$
 (3 × 3)

3. $[1 \ 2 \ 3] \begin{bmatrix} 4 \\ 5 \\ 6 \end{bmatrix} = [1 \times 4 + 2 \times 5 + 3 \times 6] = [32].$
(1 × 3) (3 × 1)   (1 × 1)

4. $\begin{bmatrix} 4 \\ 5 \\ 6 \end{bmatrix} [1 \ 2 \ 3] = \begin{bmatrix} 4 \times 1 & 4 \times 2 & 4 \times 3 \\ 5 \times 1 & 5 \times 2 & 5 \times 3 \\ 6 \times 1 & 6 \times 2 & 6 \times 3 \end{bmatrix} = \begin{bmatrix} 4 & 8 & 12 \\ 5 & 10 & 15 \\ 6 & 12 & 18 \end{bmatrix}.$
(3 × 1) (1 × 3)   (3 × 3)

5. $\begin{bmatrix} 1 & 2 & 3 \\ 4 & 5 & 6 \end{bmatrix} \begin{bmatrix} 1 \\ 1 \\ 1 \end{bmatrix} = \begin{bmatrix} 1 \times 1 + 2 \times 1 + 3 \times 1 \\ 4 \times 1 + 5 \times 1 + 6 \times 1 \end{bmatrix} = \begin{bmatrix} 6 \\ 15 \end{bmatrix}$

$\quad (2 \times 3) \ (3 \times 1) \qquad\qquad\qquad\qquad\qquad\qquad (2 \times 1)$

6. $\begin{bmatrix} 1 \\ 1 \\ 1 \end{bmatrix} \begin{bmatrix} 1 & 2 & 3 \\ 4 & 5 & 6 \end{bmatrix}$ does not exist.

$\quad (3 \times \underline{1}) \ (\underline{2} \times 3) \qquad\quad 1 \neq 2$

The results of 3 and 4 probably seem particularly strange at first; these show clearly why it is worth keeping a note of the sizes of the matrices and working out the size you expect the product to be, as we have done underneath each example.

EXERCISES 2.3

1 Let A, B, C, D be the following matrices:

$$A = \begin{bmatrix} 1 & 2 \\ 4 & -3 \end{bmatrix}, \quad B = \begin{bmatrix} 0 & 2 \\ 3 & 4 \\ 1 & 2 \end{bmatrix}, \quad C = [1 \ \ 4], \quad D = \begin{bmatrix} 4 \\ 1 \end{bmatrix}.$$

Which of the products A^2, AB, AC, AD, BA, B^2, etc. exist? Evaluate the products that do exist.

2 Evaluate AB and BA, where possible, in the following cases:

(a) $A = \begin{bmatrix} 2 & -1 & 0 & 3 \\ 4 & -5 & 1 & 0 \end{bmatrix}, \quad B = \begin{bmatrix} 0 & -3 \\ 1 & -1 \\ 2 & 1 \\ -4 & 0 \end{bmatrix};$

(b) $A = \begin{bmatrix} i & 3 & -1 \\ \frac{1}{2} & i & 1 \\ 0 & 0 & 1 \end{bmatrix}, \quad B = \begin{bmatrix} i & 1-i \\ 1 & -1 \\ 0 & 1+i \end{bmatrix};$

(c) $A = [1 \ \ 2 \ \ 3], \quad B = \begin{bmatrix} 1 \\ 2 \\ 3 \end{bmatrix};$

(d) $A = \begin{bmatrix} 1 & 0 & 0 & 0 \\ 0 & 1 & 0 & 0 \\ 0 & 0 & 2 & -1 \\ 0 & 0 & 3 & 1 \end{bmatrix}$, $B = \begin{bmatrix} 1 & 0 & 0 & 0 & 0 \\ 0 & 1 & 0 & 0 & 0 \\ 0 & 0 & \frac{3}{8} & \frac{7}{8} & 0 \\ 0 & 0 & -\frac{5}{8} & \frac{1}{8} & 0 \end{bmatrix}$.

3 Let $A = \begin{bmatrix} 1 & 2 \\ 2 & 3 \end{bmatrix}$. Show that $A^2 = 4A + I_2$, where $I_2 = \begin{bmatrix} 1 & 0 \\ 0 & 1 \end{bmatrix}$.

4 Let $A = \begin{bmatrix} \frac{1}{3} & \frac{1}{3} & \frac{1}{3} \\ \frac{1}{3} & \frac{1}{3} & \frac{1}{3} \\ \frac{1}{3} & \frac{1}{3} & \frac{1}{3} \end{bmatrix}$. Find A^n for all positive integers n.

2.4 PROPERTIES OF MATRIX MULTIPLICATION

Let A, B, C be any real or complex matrices, and let α be any real or complex number. When all the following sums and products are defined, matrix multiplication satisfies the following properties:

M1 $(AB)C = A(BC)$.

M2 $A(B + C) = AB + AC$.

M3 $(A + B)C = AC + BC$.

M4 $\alpha(AB) = (\alpha A)B = A(\alpha B)$.

Implicit in all of these is that if either side is defined then so is the other, and equality then results. Property M1 is known as the *associative law of multiplication*; M2 and M3 are usually called the *distributive laws of multiplication over addition*. One unfortunate outcome of the more complicated product we have adopted is that M1, M2 and M3 are not quite so straightforward to check. Consequently, we will prove M1 and M2; M3 is similar to M2, and M4 is easy.

We will refer to the element in the ith row and jth column of a matrix as its (i,j)-*element*.

Proof of M1

Let

$A = [a_{ij}]_{m \times n}$, $B = [b_{ij}]_{n \times p}$, $C = [c_{ij}]_{p \times q}$, $D = AB = [d_{ij}]_{m \times p}$, $E = BC = [e_{ij}]_{n \times q}$.

Then the (i,j)-element of DC is

$$\sum_{k=1}^{p} d_{ik} c_{kj} = \sum_{k=1}^{p} \left(\sum_{r=1}^{n} a_{ir} b_{rk} \right) c_{kj},$$

and the (i,j)-element of AE is

$$\sum_{r=1}^{n} a_{ir}e_{rj} = \sum_{r=1}^{n} a_{ir}\left(\sum_{k=1}^{p} b_{rk}c_{kj}\right).$$

But these are the same, so that $(AB)C = A(BC)$.

Proof of M2

Let

$A = [a_{ij}]_{m \times n}, B = [b_{ij}]_{n \times p}, C = [c_{ij}]_{n \times p}, D = AB = [d_{ij}]_{m \times p}, E = AC = [e_{ij}]_{m \times p}.$

Then $B + C = [b_{ij} + c_{ij}]_{n \times p}$,
the (i,j)-element of $A(B + C)$ is

$$\sum_{k=1}^{n} a_{ik}(b_{kj} + c_{kj}),$$

the (i,j)-element of $AB + AC$ is

$$d_{ij} + e_{ij} = \sum_{k=1}^{n} a_{ik}b_{kj} + \sum_{k=1}^{n} a_{ik}c_{kj}.$$

Again these are equal, and we have $A(B + C) = AB + AC$.

It follows from M1 that arbitrary finite products are unambiguous without bracketing; a straightforward, though slightly tedious, induction argument is all that is required to establish this. In particular, we can define (provided that A is square)

$$A^n = \underbrace{AA\ldots A}_{n \text{ terms}} \quad \text{for all } n \in \mathbb{Z}^+,$$

and it is clear that

$$A^m A^n = A^n A^m = A^{m+n}, (A^m)^n = (A^n)^m = A^{mn} \quad \text{for all } m, n \in \mathbb{Z}^+.$$

Likewise, the distributive laws can also be generalised to

$$A(\alpha_1 B_1 + \ldots + \alpha_n B_n) = \alpha_1 AB_1 + \ldots + \alpha_n AB_n.$$

The square matrix in which the elements on the main diagonal are 1's and all other elements are 0 is called the $(n \times n)$ *identity matrix*. We denote it by I, or by I_n if we want to emphasise its size. Then the following two laws are easy to check:

M5   If A is an $n \times p$ matrix, then $AI_p = I_n A = A$.

M6   if A is an $n \times p$ matrix, then $O_{m \times n} A = O_{m \times p}$ and $AO_{p \times q} = O_{n \times q}$.

Examples 2.4

1. $(A + B)^2 = (A + B)(A + B) = A(A + B) + B(A + B) = A^2 + AB + BA + B^2$.

2. $(A+B)(A-B) = A(A-B) + B(A-B) = A^2 - AB + BA - B^2$.

3. Find all 2×2 complex matrices A such that $A^2 = I_2$.

Solution Let $A = \begin{bmatrix} a & b \\ c & d \end{bmatrix}$. Then $A^2 = \begin{bmatrix} a^2 + bc & ab + bd \\ ca + dc & cb + d^2 \end{bmatrix}$,

so we require
$$a^2 + bc = cb + d^2 = 1 \qquad (1)$$
$$ab + bd = ca + dc = 0. \qquad (2)$$

From equation (2), either $b = c = 0$ or else $a = -d$. By substituting into (1) we see that if $b = c = 0$ then $a = \pm 1$, $d = \pm 1$, and if $a = -d$ then $a = \pm\sqrt{1-bc}$. Thus A is of the form

$$\begin{bmatrix} \pm\sqrt{1-bc} & b \\ c & \mp\sqrt{1-bc} \end{bmatrix} \quad \text{or} \quad \begin{bmatrix} \pm 1 & 0 \\ 0 & \pm 1 \end{bmatrix}.$$

It is very tempting to write the answer to example 1 above as $A^2 + 2AB + B^2$, and to example 2 as $A^2 - B^2$. However, this requires that $AB = BA$ and, in general, *matrix multiplication is not commutative*. For example, let

$$A = \begin{bmatrix} 1 & 0 \\ 0 & 0 \end{bmatrix}, \qquad B = \begin{bmatrix} 0 & 0 \\ 1 & 0 \end{bmatrix}.$$

Then $AB = \begin{bmatrix} 0 & 0 \\ 0 & 0 \end{bmatrix}$, whereas $BA = \begin{bmatrix} 0 & 0 \\ 1 & 0 \end{bmatrix}$.

This does not mean that AB is *never* the same as BA: we simply cannot assert that AB is *always* the same as BA. These same two matrices illustrate another curious phenomenon; we have $AB = O$, but $A \neq O$ and $B \neq O$.

Matrices A, B such that $AB = O$ are called *zero divisors*. Clearly, O itself is a zero divisor, albeit a not very significant one! Consequently, if $A, B \neq O$ but $AB = O$, we refer to A, B as *proper* zero divisors.

Another familiar property of numbers which fails for matrices is the *cancellation law*. From the fact that $AB = AC$, or that $BA = CA$, we *cannot* deduce that $B = C$ necessarily.

EXERCISES 2.4

1 Let $A = \begin{bmatrix} 1 & 1 & 0 \\ 0 & 0 & 1 \\ 0 & 0 & -1 \end{bmatrix}$. Calculate A^2, A^3 and hence find A^n for every $n \in \mathbb{Z}^+$.

2 Prove properties M3 to M6.

3 Let $A = \begin{bmatrix} 1 & 0 & 0 \\ 1 & 1 & 0 \\ 0 & 1 & 1 \end{bmatrix}$, and let B_1, B_2 be 3×3 real matrices such that $B_1 A = AB_1$, $B_2 A = AB_2$. Prove that $B_1 B_2 = B_2 B_1$.

4 Find all 2×2 complex matrices A such that $A^2 = -I_2$.

5 Find three 2×2 real matrices A, B, C such that $AB = AC$ but $B \neq C$.

6 Find two 2×2 real matrices A, B such that $(AB)^2 \neq A^2 B^2$.

7 Let $A = [a_{ij}]_{m \times n}$, $B = [b_{ij}]_{n \times p}$, and suppose that every element of row r of A is zero. Prove that every element of row r of AB is zero. State and prove a corresponding result for columns.

8 Let A be a 2×2 real matrix such that $AB = BA$ for all 2×2 real matrices B. Show that $A = \lambda I_2$ for some $\lambda \in \mathbb{R}$.

2.5 THE TRANSPOSE OF A MATRIX

The transpose, A^T, of an $m \times n$ matrix A is the $n \times m$ matrix obtained by interchanging the rows and columns of A; that is, if $A = [a_{ij}]_{m \times n}$, then $A^T = [a_{ji}]_{n \times m}$.

Examples 2.5.1

1. If $A = \begin{bmatrix} 1 & 2 \\ 3 & 4 \end{bmatrix}$ then $A^T = \begin{bmatrix} 1 & 3 \\ 2 & 4 \end{bmatrix}$.

2. If $A = \begin{bmatrix} 1 & -1 \end{bmatrix}$ then $A^T = \begin{bmatrix} 1 \\ -1 \end{bmatrix}$.

Let A, B be real or complex matrices, and let α be any real or complex number. Then the transpose satisfies the following properties:

T1 $(A + B)^T = A^T + B^T$.

T2 $(\alpha A)^T = \alpha A^T$.

T3 $(AB)^T = B^T A^T$. (*Note: Not* $A^T B^T$.)

T4 $(A^T)^T = A$.

Again these are to be interpreted as meaning that if either side is defined then so is the other, and equality results. As usual, T1 and T3 can be generalised by a simple induction proof to any finite sum or product.

Properties T1, T2 and T4 are easy to check and are left as exercises; we will prove T3.

Proof of T3

Let

$$A = [a_{ij}]_{m \times n}, \quad B = [b_{ij}]_{n \times p}, \quad C = A^T = [c_{ij}]_{n \times m} \quad (\text{so } c_{ij} = a_{ji}),$$

$$D = B^T = [d_{ij}]_{p \times n} \text{ (so } d_{ij} = b_{ji}), \quad E = AB = [e_{ij}]_{m \times p} \left(\text{so } e_{ij} = \sum_{k=1}^{n} a_{ik} b_{kj}\right).$$

Then the (i,j)-element of $(AB)^T$ = the (j,i)-element of $AB = e_{ji} = \sum_{k=1}^{n} a_{jk} b_{ki}$, and the (i,j)-element of $B^T A^T = \sum_{k=1}^{n} d_{ik} c_{kj} = \sum_{k=1}^{n} b_{ki} a_{jk} = \sum_{k=1}^{n} a_{jk} b_{ki}$.

Since $(AB)^T$ and $B^T A^T$ are both $p \times m$ matrices and their (i,j)-elements are equal, they are themselves equal.

A square matrix A for which $A = A^T$ is called *symmetric*; if $A^T = -A$ then A is termed *antisymmetric*. If $A = [a_{ij}]_{n \times n}$ is symmetric, then $a_{ij} = a_{ji}$ for all $1 \leq i, j \leq n$, so that the array of elements of A is symmetric about the main diagonal. If $A = [a_{ij}]_{n \times n}$ is antisymmetric, then $a_{ij} = -a_{ji}$ for all $1 \leq i, j \leq n$; when $i = j$ this implies that $a_{ii} = 0$ for $1 \leq i \leq n$. It follows that the elements on the main diagonal of an antisymmetric matrix are all 0, and the rest of the array of elements is antisymmetric (in the sense that $a_{ij} = -a_{ji}$) about the main diagonal.

Examples 2.5.2

1. If A is any $n \times n$ matrix, then
$$(A + A^T)^T = A^T + (A^T)^T = A^T + A = A + A^T,$$
so that $A + A^T$ is symmetric.

2. If A is any $n \times n$ matrix, then
$$(A - A^T)^T = A^T - (A^T)^T = A^T - A = -(A - A^T),$$
so that $A - A^T$ is antisymmetric.

3. If A is any $n \times n$ matrix, then
$$A = \tfrac{1}{2}(A + A^T) + \tfrac{1}{2}(A - A^T),$$
so that A may be written as a sum of a symmetric and an antisymmetric matrix.

GUIDE TO LINEAR ALGEBRA

EXERCISES 2.5

1. Write down the transposes of the following matrices:

   (a) $\begin{bmatrix} 1 & 2 \\ 3 & 4 \\ 5 & 6 \end{bmatrix}$, (b) $\begin{bmatrix} 3 & 4 & -1 \\ 0 & -1 & 2 \\ 8 & 1 & 4 \end{bmatrix}$, (c) $\begin{bmatrix} 1 & -1 \\ 0 & 1 \end{bmatrix}$,

   (d) $[1 \ \ 3 \ \ 4]$.

2. Find all 2×2 real matrices A such that $A^T = 2A$.

3. Use example 2.5.2.3 above to write each of the following matrices as a sum of a symmetric matrix and an antisymmetric matrix:

   (a) $\begin{bmatrix} 1 & 2 \\ 3 & 4 \end{bmatrix}$, (b) $\begin{bmatrix} 1 & 1 \\ 1 & 1 \end{bmatrix}$, (c) $\begin{bmatrix} 1 & -1 & 3 \\ 4 & 2 & -1 \\ -2 & 1 & 2 \end{bmatrix}$.

4. Are there any matrices which are both symmetric and antisymmetric?

5. Prove that if A is any $n \times n$ matrix then AA^T is symmetric.

6. Find two 2×2 real matrices A, B such that $(AB)^T \neq A^T B^T$.

7. For each of the statements listed below give either a proof or a counter-example.
   (a) $AA^T = A^T A$ for every matrix A.
   (b) If A, B are symmetric matrices such that $AB = BA$, then AB is symmetric.
   (c) If A is a symmetric matrix, then so is A^n for each $n \in \mathbb{Z}^+$.
   (d) If A, B are $n \times n$ symmetric matrices, then so is AB.

8. Let A be a square matrix, and suppose that $A = B + C$ where B is symmetric and C is antisymmetric. Show that
$$B = \tfrac{1}{2}(A + A^T), \qquad C = \tfrac{1}{2}(A - A^T).$$

9. Suppose that A, B are matrices such that $AB = A$, $BA = B$. Show that
$$(A^T)^2 = A^T, \qquad (B^T)^2 = B^T.$$

2.6 INVERTIBLE MATRICES

A square matrix A is said to be *invertible* if there is a matrix B such that $AB = BA = I$; in this case, B is called an *inverse* of A. It is clear that B must also be square: in fact, it has the same size as A.

Examples 2.6.1

1. I_n has inverse I_n.

2. $-I_n$ has inverse $-I_n$.

3. The matrices $\begin{bmatrix} 0 & 0 \\ 0 & 0 \end{bmatrix}, \begin{bmatrix} 1 & 0 \\ 0 & 0 \end{bmatrix}, \begin{bmatrix} 1 & 1 \\ 0 & 0 \end{bmatrix}, \begin{bmatrix} 1 & 1 \\ 1 & 1 \end{bmatrix}$, for example, have no inverses. We will show this for the last of these matrices; the others can be checked similarly.

Suppose that
$$\begin{bmatrix} a+c & b+d \\ a+c & b+d \end{bmatrix} = \begin{bmatrix} 1 & 1 \\ 1 & 1 \end{bmatrix} \begin{bmatrix} a & b \\ c & d \end{bmatrix} = \begin{bmatrix} 1 & 0 \\ 0 & 1 \end{bmatrix}.$$

Then
$$a+c=1 \quad \text{and} \quad a+c=0,$$
$$b+d=0 \quad \text{and} \quad b+d=1.$$

Clearly there is no possible solution to these equations.

THEOREM 2.6.1 A square matrix has at most one inverse.

Proof Suppose that the $n \times n$ matrix A has inverses B, C, so that
$$AB = BA = I \quad \text{and} \quad AC = CA = I.$$
Then $\qquad B = IB = (CA)B = C(AB) = CI = C.$

If A is invertible, it is usual to denote the unique inverse of A by A^{-1}. Let A, B be invertible real or complex matrices, and let α be any non-zero real or complex number. Then the following properties hold:

I1 αA is invertible and $(\alpha A)^{-1} = \alpha^{-1} A^{-1}$.

I2 AB is invertible (if defined) and $(AB)^{-1} = B^{-1}A^{-1}$. (*Note:* Not $A^{-1}B^{-1}$.)

I3 A^{-1} is invertible and $(A^{-1})^{-1} = A$.

I4 A^T is invertible and $(A^T)^{-1} = (A^{-1})^T$.

The essential idea required to prove all of these is that in order to show that B is the inverse of A it suffices to check that $AB = BA = I$.

Proof of I1 $(AB)(B^{-1}A^{-1}) = ((AB)B^{-1})A^{-1} = (A(BB^{-1}))A^{-1} = (AI)A^{-1} = AA^{-1} = I$. Similarly, $(B^{-1}A^{-1})(AB) = I$, from which the result follows.

Proof of I2 $(\alpha A)(\alpha^{-1}A^{-1}) = (\alpha \alpha^{-1})AA^{-1} = 1I = I$. Similarly, $(\alpha^{-1}A^{-1})(\alpha A) = I$ so that $(\alpha A)^{-1} = \alpha^{-1}A^{-1}$.

Proof of I3 The equations $AA^{-1} = A^{-1}A = I$ show that $(A^{-1})^{-1}$ exists and is equal to A.

Proof of I4 We have $AA^{-1} = A^{-1}A = I$. Taking transposes gives $(AA^{-1})^T = (A^{-1}A)^T = I^T$. Hence $(A^{-1})^T A^T = A^T(A^{-1})^T = I$, from which it follows that A^T is invertible and $(A^T)^{-1} = (A^{-1})^T$.

It is straightforward to generalise I2 by means of an induction argument to show that if $A_1, \ldots, A_n$ are invertible matrices of the same size, then $A_1 A_2 \ldots A_n$ is invertible and

$$(A_1 A_2 \ldots A_n)^{-1} = A_n^{-1} \ldots A_2^{-1} A_1^{-1}.$$

If A is invertible and n is a negative integer, we can now define

$$A^n = (A^{-1})^{-n} = A^{-1} A^{-1} \ldots A^{-1} \quad (-n \text{ terms}).$$

To complete the sequence we define $A^0 = I$. The index laws

$$A^m A^n = A^n A^m = A^{m+n}, \quad (A^m)^n = (A^n)^m = A^{nm}$$

can now be checked for every $m, n \in \mathbb{Z}$.

Two questions arise naturally at this point:

(a) Is there a practical method for finding the inverse of an invertible matrix?

(b) Is there a practical criterion for determining whether a given square matrix is invertible?

We will address these questions in the next two chapters.

Example 2.6.2

Let A be an $n \times n$ matrix such that $A^2 + 2A + I = O$. Prove that A is invertible and find its inverse.

Solution We employ the strategy used in establishing properties I1 to I4 above. The equation satisfied by A can be rewritten as

$$-A^2 - 2A = I.$$

But then, factorising the left-hand side yields

$$A(-A - 2I) = (-A - 2I)A = I.$$

From this it is clear that A is invertible with inverse $A^{-1} = -A - 2I$.

EXERCISES 2.6

1 Show that the following matrices are not invertible:

(a) $\begin{bmatrix} 2 & 3 \\ 0 & 0 \end{bmatrix}$, (b) $\begin{bmatrix} 1 & 2 \\ 2 & 4 \end{bmatrix}$, (c) $\begin{bmatrix} 1 & 1 & 1 \\ 0 & 1 & 1 \\ 0 & 0 & 0 \end{bmatrix}$.

2 Let A be an $n \times n$ matrix satisfying $A^3 - 2A^2 + 5A - 4I = O$. Show that A is invertible and find A^{-1}. Does a generalisation suggest itself?

3 (a) Suppose that A is an $n \times n$ matrix, that I is the $n \times n$ identity matrix, and that k is a positive integer. Prove that

$$(I - A^{k+1}) = (I - A)(I + A + A^2 + \ldots + A^k).$$

Now suppose that $A^{k+1} = O$. Prove that $I - A$ is invertible, and express $(I - A)^{-1}$ as a polynomial in I and A.

(b) Let $B = \begin{bmatrix} 5 & -3 & -1 \\ 4 & -2 & -1 \\ 6 & -5 & 0 \end{bmatrix}$.

Calculate $(I - B)$, $(I - B)^2$ and $(I - B)^3$. By writing $B = I - (I - B)$, prove that B is invertible and calculate B^{-1}.

4 The square matrix A is called a *scalar* matrix if $A = \lambda I$ for some real or complex number λ. Show that the scalar matrix $A = \lambda I$ is invertible if and only if $\lambda \neq 0$, and find A^{-1} when it exists.

5 The square matrix $A = [a_{ij}]_{n \times n}$ is called a *diagonal* matrix if $a_{ij} = 0$ whenever $i \neq j$; we write $A = \text{diag}(a_{11}, a_{22}, \ldots, a_{nn})$. Determine necessary and sufficient conditions on $a_{11}, a_{22}, \ldots, a_{nn}$ for such a diagonal matrix to be invertible, and, when it is, describe its inverse.

6 The matrix A is called *nilpotent* if there is a positive integer k such that $A^k = O$. Show that nilpotent matrices cannot be invertible.

7 Find two 2×2 real invertible matrices A, B such that $(AB)^{-1} \neq A^{-1}B^{-1}$.

8 Let A be an $n \times n$ matrix for which there is exactly one matrix B such that $AB = I$. By considering $BA + B - I$ show that A is invertible.

9 Let $A + I$ be invertible. Show that $(A + I)^{-1}$ and $(I - A)$ commute (that is, $(A + I)^{-1}(I - A) = (I - A)(A + I)^{-1}$).

SOLUTIONS AND HINTS FOR EXERCISES

Exercises 2.1

1 (a) 4×2; (b) 1×6; (c) 3×3; (d) 2×2.

2 (c) is square—the main diagonal is $1, -1, 3$;
   (d) is square—the main diagonal is $1, 1$.

3 $\begin{bmatrix} 3 \\ -1 \\ 1 \end{bmatrix}$. **4** $[6 \quad 1 \quad 3]$.

Exercises 2.2

1 $\begin{bmatrix} 3 \\ 1 \end{bmatrix}$ **2** $[5 \quad -4]$ **3** not defined   **4** $\begin{bmatrix} -3+2i & 3+3i \\ -i & i \end{bmatrix}$

5 not defined   **6** $\begin{bmatrix} -1 & 0 \\ -1 & -5 \\ 1 & -1 \end{bmatrix}$ **7** $\begin{bmatrix} -3 & -3i \\ 3+12i & \frac{3}{2}i \end{bmatrix}$

8 $\begin{bmatrix} 1 & -1 \\ 2 & 3 \end{bmatrix}$ **9** $\begin{bmatrix} 1 & 8 & -5 & 0 \\ 24 & 14 & 17 & -3 \\ -5 & 21 & -6 & 0 \\ -12 & 12 & -2 & 19 \end{bmatrix}$

10 not defined   **11** $[-2-i \quad 2i]$ **12** $\begin{bmatrix} 3 \\ 4 \end{bmatrix}$

Exercises 2.3

1 $A^2 = \begin{bmatrix} 9 & -4 \\ -8 & 17 \end{bmatrix}$, $AD = \begin{bmatrix} 6 \\ 13 \end{bmatrix}$, $BA = \begin{bmatrix} 8 & -6 \\ 19 & -6 \\ 9 & -4 \end{bmatrix}$,

$BD = \begin{bmatrix} 2 \\ 16 \\ 6 \end{bmatrix}$, $CA = [17 \quad -10]$, $CD = [8]$, $DC = \begin{bmatrix} 4 & 16 \\ 1 & 4 \end{bmatrix}$.

No other products exist.

2 (a) $AB = \begin{bmatrix} -13 & -5 \\ -3 & -6 \end{bmatrix}$,

$BA = \begin{bmatrix} -12 & 15 & -3 & 0 \\ -2 & 4 & -1 & 3 \\ 8 & -7 & 1 & 6 \\ -8 & 4 & 0 & -12 \end{bmatrix}$;

(b) $AB = \begin{bmatrix} 2 & -3 \\ \frac{3}{2}i & \frac{3}{2} - \frac{1}{2}i \\ 0 & 1+i \end{bmatrix}$, BA does not exist;

(c) $AB = [14]$, $BA = \begin{bmatrix} 1 & 2 & 3 \\ 2 & 4 & 6 \\ 3 & 6 & 9 \end{bmatrix}$;

(d) $AB = \begin{bmatrix} 1 & 0 & 0 & 0 & 0 \\ 0 & 1 & 0 & 0 & 0 \\ 0 & 0 & \frac{11}{8} & \frac{13}{8} & 0 \\ 0 & 0 & \frac{1}{2} & \frac{11}{4} & 0 \end{bmatrix}$, BA does not exist.

3 This is straightforward.

4 It is easily seen that $A^n = A$ for all n.

Exercises 2.4

1 $A^2 = \begin{bmatrix} 1 & 1 & 1 \\ 0 & 0 & -1 \\ 0 & 0 & 1 \end{bmatrix}$, $A^3 = A$. A straightforward induction argument then shows that $A^n = A$ if n is odd, and $A^n = A^2$ if n is even.

2 M3: Let $A = [a_{ij}]_{m \times n}$, $B = [b_{ij}]_{m \times n}$, $C = [c_{ij}]_{n \times p}$, $D = AC = [d_{ij}]_{m \times p}$, $E = BC = [e_{ij}]_{m \times p}$. Then $A + B = [a_{ij} + b_{ij}]_{m \times n}$, the (i,j)-element of $(A+B)C$ is $\sum_{k=1}^{n}(a_{ik} + b_{ik})c_{kj}$, and the (i,j)-element of $AC + BC$ is $d_{ij} + e_{ij} = \sum_{k=1}^{n} a_{ik}c_{kj} + \sum_{k=1}^{n} b_{ik}c_{kj}$. The result is now clear.

M4: Let $A = [a_{ij}]_{m \times n}$, $B = [b_{ij}]_{n \times p}$.
Then the (i,j)-element of $\alpha(AB)$ is $\alpha \sum_{k=1}^{n} a_{ik}b_{kj}$,
the (i,j)-element of $(\alpha A)B$ is $\sum_{k=1}^{n} (\alpha a_{ik})b_{kj}$,
the (i,j)-element of $A(\alpha B)$ is $\sum_{k=1}^{n} a_{ik}(\alpha b_{kj})$; these are clearly equal.

M5: Let $A = [a_{ij}]_{n \times p}$. Then $I = [\delta_{ij}]$ where

$$\delta_{ij} = \begin{cases} 0 & \text{if } i \neq j \\ 1 & \text{if } i = j \end{cases},$$

so the (i,j)-element of AI_p is $\sum_{k=1}^{p} a_{ik}\delta_{kj} = a_{ij}$,

and the (i,j)-element of $I_n A$ is $\sum_{k=1}^{n} \delta_{ik}a_{kj} = a_{ij}$,

from which the result is clear.

M6: This is straightforward.

3 Putting $B = [b_{ij}]_{3 \times 3}$, multiplying out AB and BA, and equating corresponding elements shows that $AB = BA$ if and only if B has the form $\begin{bmatrix} a & 0 & 0 \\ b & a & 0 \\ c & b & a \end{bmatrix}$. Putting $B_1 = \begin{bmatrix} a_1 & 0 & 0 \\ b_1 & a_1 & 0 \\ c_1 & b_1 & a_1 \end{bmatrix}$, $B_2 = \begin{bmatrix} a_2 & 0 & 0 \\ b_2 & a_2 & 0 \\ c_2 & b_2 & a_2 \end{bmatrix}$, the result is now easy to check.

4 Let $A = \begin{bmatrix} a & b \\ c & d \end{bmatrix}$. Then $\begin{bmatrix} a^2 + bc & ab + bd \\ ca + dc & cb + d^2 \end{bmatrix} = \begin{bmatrix} -1 & 0 \\ 0 & -1 \end{bmatrix}$ implies that $a^2 + bc = cb + d^2 = -1$, $c(a + d) = b(a + d) = 0$.
If $a + d = 0$ then $a = \pm i\sqrt{1 + bc}$; if $a + d \neq 0$ then $b = c = 0$, $a = \pm i$, $d = \pm i$.
Hence A has the form

$$\begin{bmatrix} \pm i\sqrt{1 + bc} & b \\ c & \mp i\sqrt{1 + bc} \end{bmatrix}, \quad \text{or} \quad \begin{bmatrix} \pm i & 0 \\ 0 & \pm i \end{bmatrix} (b, c \in \mathbb{C}).$$

5 $A = B = \begin{bmatrix} 1 & 0 \\ 0 & 0 \end{bmatrix}$, $C = \begin{bmatrix} 1 & 0 \\ 1 & 0 \end{bmatrix}$ will do.

6 $A = \begin{bmatrix} 0 & 1 \\ 0 & 0 \end{bmatrix}$, $B = \begin{bmatrix} 0 & 0 \\ 1 & 0 \end{bmatrix}$ for example.

7 We are given that $a_{rj} = 0$ for $1 \leq j \leq n$. A typical element of row r of AB is the (r, j)-element which is $\sum_{k=1}^{n} a_{rk} b_{kj} = 0$. Similarly, if every element of column s of B is zero, then every element of column s of AB is zero.

8 If $A = \begin{bmatrix} a & b \\ c & d \end{bmatrix}$, $B = \begin{bmatrix} e & f \\ g & h \end{bmatrix}$,

then $AB = \begin{bmatrix} ae + bg & af + bh \\ ce + dg & cf + dh \end{bmatrix}$, $BA = \begin{bmatrix} ae + fc & eb + fd \\ ga + hc & gb + hd \end{bmatrix}$.

Putting $AB = BA$ gives $bg = fc, f(a - d) = b(e - h), c(e - h) = g(a - d)$. We want this equation to hold for all 2×2 real matrices B, so, in particular, picking $f = g = 0$, $e \neq h$ we have $b = c = 0$; similarly, picking $f = g = 1$ we get $a = d = \lambda$, say.

Exercises 2.5

1 (a) $\begin{bmatrix} 1 & 3 & 5 \\ 2 & 4 & 6 \end{bmatrix}$; (b) $\begin{bmatrix} 3 & 0 & 8 \\ 4 & -1 & 1 \\ -1 & 2 & 4 \end{bmatrix}$; (c) $\begin{bmatrix} 1 & 0 \\ -1 & 1 \end{bmatrix}$;

(d) $\begin{bmatrix} 1 \\ 3 \\ 4 \end{bmatrix}$.

2 Proceed as in question 4 of Exercises 2.4 (only this one is easier), and you should discover that $A = O$ is the only possibility.

3 This is straightforward.

4 Clearly, $A^T = A = -A$ implies that $2A = O$, so $A = O$ is the only possibility.

5 Note that $(AA^T)^T = (A^T)^T A^T = AA^T$, whence the result.

6 $A = \begin{bmatrix} 0 & 1 \\ 0 & 0 \end{bmatrix}$, $B = \begin{bmatrix} 0 & 0 \\ 1 & 0 \end{bmatrix}$ for example.

7 (a) False: Let $A = \begin{bmatrix} 0 & 1 \\ 0 & 0 \end{bmatrix}$ for instance.
  (b) True: $(AB)^T = (BA)^T = A^T B^T = AB$.
  (c) True: Use (b) above and induction on n.
  (d) False: Let $A = \begin{bmatrix} 1 & 0 \\ 0 & 0 \end{bmatrix}$, $B = \begin{bmatrix} 0 & 1 \\ 1 & 0 \end{bmatrix}$ for example.

8 We have
$$A = B + C \tag{1}$$
and
$$A^T = (B + C)^T = B^T + C^T = B - C. \tag{2}$$
Adding (1) and (2) gives $A + A^T = 2B$;
subtracting (2) from (1) gives $A - A^T = 2C$.

9 We have $A^T = (AB)^T = B^T A^T$, and $B^T = (BA)^T = A^T B^T$.

Hence
$$(A^T)^2 = A^T A^T = (B^T A^T)(B^T A^T) = B^T(A^T B^T)A^T = B^T B^T A^T = B^T(B^T A^T)$$
$$= B^T A^T = A^T,$$
and
$$(B^T)^2 = B^T B^T = (A^T B^T)(A^T B^T) = A^T(B^T A^T)B^T = A^T A^T B^T = A^T(A^T B^T)$$
$$= A^T B^T = B^T.$$

Exercises 2.6

1 (a) $\begin{bmatrix} 2 & 3 \\ 0 & 0 \end{bmatrix} \begin{bmatrix} a & b \\ c & d \end{bmatrix} = \begin{bmatrix} 1 & 0 \\ 0 & 1 \end{bmatrix} \Rightarrow \begin{array}{l} 2a + 3c = 1, \; 2b + 3d = 0. \\ 0 = 0, \qquad 0 = 1! \end{array}$

(b) $\begin{bmatrix} 1 & 2 \\ 2 & 4 \end{bmatrix} \begin{bmatrix} a & b \\ c & d \end{bmatrix} = \begin{bmatrix} 1 & 0 \\ 0 & 1 \end{bmatrix} \Rightarrow \begin{array}{l} a + 2c = 1, \; b + 2d = 0 \\ 2a + 4c = 0, \; 2b + 4d = 1 \end{array} \Rightarrow 2 = 0!$

(c) $\begin{bmatrix} 1 & 1 & 1 \\ 0 & 1 & 1 \\ 0 & 0 & 0 \end{bmatrix} \begin{bmatrix} a & b & c \\ d & e & f \\ g & h & i \end{bmatrix} = \begin{bmatrix} 1 & 0 & 0 \\ 0 & 1 & 0 \\ 0 & 0 & 1 \end{bmatrix}$

$\begin{array}{lll} a+d+g = 1, & b+e+h = 0, & c+f+i = 0 \\ \Rightarrow \quad d+g = 0, & e+h = 1, & f+i = 0 \\ 0 = 0, & 0 = 0, & 0 = 1! \end{array}$

2 $0 = A^3 - 2A^2 + 5A - 4I$; $\quad I = \frac{1}{4}(A^3 - 2A^2 + 5A) = \frac{1}{4}(A^2 - 2A + 5I)A$;
$A^{-1} = \frac{1}{4}(A^2 - 2A + 5I)$.

Suppose that A satisfies a polynomial equation such as
$$a_k A^k + a_{k-1} A^{k-1} + \ldots + a_1 A + a_0 I = O.$$
Then $\quad a_0 I = -A(a_k A^{k-1} + \ldots + a_1 I).$

Hence, if $\quad a_0 \neq 0, \quad A^{-1} = -a_0^{-1}(a_k A^{k-1} + \ldots + a_1 I).$

3 (a) In order to establish the first equation simply expand the right-hand side.
Now $A^{k+1} = O \Rightarrow I = (I - A)(I + A + \ldots + A^k)$
$$\Rightarrow (I - A)^{-1} = I + A + A^2 + \ldots + A^k.$$

(b) $I - B = \begin{bmatrix} -4 & 3 & 1 \\ -4 & 3 & 1 \\ -6 & 5 & 1 \end{bmatrix}$, $(I - B)^2 = \begin{bmatrix} -2 & 2 & 0 \\ -2 & 2 & 0 \\ -2 & 2 & 0 \end{bmatrix}$,

$(I - B)^3 = \begin{bmatrix} 0 & 0 & 0 \\ 0 & 0 & 0 \\ 0 & 0 & 0 \end{bmatrix}$.

Putting $A = I - B$, $k = 2$, in (a), we see that $B = I - (I - B) = I - A$ is invertible and $B^{-1} = I + (I - B) + (I - B)^2 = \begin{bmatrix} -5 & 5 & 1 \\ -6 & 6 & 1 \\ -8 & 7 & 2 \end{bmatrix}$.

4 $A = \lambda I$ is invertible $\Rightarrow A^{-1}$ exists such that $I = AA^{-1} = \lambda I A^{-1} = \lambda A^{-1}$
$$\Rightarrow \lambda \neq 0 \text{ and } A^{-1} = \lambda^{-1} I.$$

5 If A^{-1} exists, let $A^{-1} = [b_{ij}]_{n \times n}$. Then
A is invertible $\Rightarrow A^{-1}$ exists such that $I = AA^{-1} = \left[\sum_{k=1}^{n} a_{ik} b_{kj}\right]_{n \times n}$
$\Rightarrow \sum_{k=1}^{n} a_{ik} b_{kj} = a_{ii} b_{ij} = \delta_{ij}$
$\Rightarrow a_{ii} b_{ii} = 1 \quad$ and $\quad a_{ii} b_{ij} = 0 \quad$ if $i \neq j \quad (1 \leq i \leq n)$
$\Rightarrow a_{ii} \neq 0, \quad b_{ii} = a_{ii}^{-1}, \quad b_{ij} = 0 \quad$ if $i \neq j \quad (1 \leq i \leq n)$.

Hence, A is invertible if and only if $a_{ii} \neq 0$ for each $1 \leq i \leq n$; in this case $A^{-1} = \mathrm{diag}(a_{11}^{-1}, a_{22}^{-1}, \ldots, a_{nn}^{-1})$.

6 Suppose that A is nilpotent and is invertible. Then $A^k = O$ and there is an A^{-1} such that $AA^{-1} = I$. Clearly $k > 1$, since $O = OA^{-1} = I$ is impossible. Let r be the smallest positive integer such that $A^r = O$, so that $A^{r-1} \neq O$. Then
$$A^{r-1} = A^{r-1} I = A^{r-1}(AA^{-1}) = A^r A^{-1} = OA = O, \quad \text{a contradiction.}$$

7 $A = \begin{bmatrix} 0 & 1 \\ 1 & 0 \end{bmatrix}$, $B = \begin{bmatrix} 0 & 2 \\ 1 & 0 \end{bmatrix}$ will suffice.

8 Note that $A(BA + B - I) = ABA + AB - A = IA + I - A = I$. The unique-

ness of B now implies that $BA + B - I = B$, and hence that $BA = I$. Thus A is invertible and $A^{-1} = B$.

9 We have $\qquad (A+I)(A-I) = A^2 - A + A - I = A^2 - I,$

and $\qquad (A-I)(A+I) = A^2 - A + A - I = A^2 - I,$

so $\qquad (A+I)(A-I) = (A-I)(A+I).$

Multiplying both sides of this equation on both the right and the left by $(A+I)^{-1}$ gives the required result.

3 ROW REDUCTION

3.1 SYSTEMS OF LINEAR EQUATIONS

A *linear equation* in n unknowns $x_1, \ldots, x_n$ is an equation of the form
$$a_1 x_1 + a_2 x_2 + \ldots + a_n x_n = b,$$
where $a_1, a_2, \ldots, a_n, b$ are real or complex numbers, and are referred to as the *coefficients* of the equation. When $n = 2$ and a_1, a_2 are real, such an equation is the Cartesian equation of a straight line in $\mathbb{R}^2$ (Fig. 3.1(a)); when $n = 3$ and a_1, a_2, a_3 are real we have the Cartesian equation of a plane in $\mathbb{R}^3$ (Fig. 3.1(b)); see section 1.5(f).

When $n > 3$ such a pictorial representation is no longer possible, but, by analogy with the cases $n = 2, 3$, geometers speak of such an equation as representing a *hyperplane*.

Now consider a system of linear equations
$$a_{11} x_1 + a_{12} x_2 + \ldots + a_{1n} x_n = b_1$$
$$a_{21} x_1 + a_{22} x_2 + \ldots + a_{2n} x_n = b_2$$
$$\ldots \ldots \ldots \ldots \ldots \ldots \ldots \ldots \ldots \ldots \ldots$$
$$a_{m1} x_1 + a_{m2} x_2 + \ldots + a_{mn} x_n = b_m$$

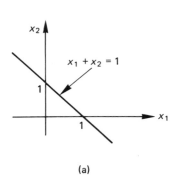
(a)

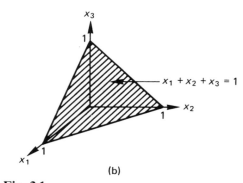
(b)

Fig. 3.1

in $x_1, x_2, \ldots, x_n$ where the a_{ij}s, b_is are real or complex numbers. A sequence of values $x_1 = c_1, x_2 = c_2, \ldots, x_n = c_n$ (c_is real or complex) satisfying all of the equations is called a *solution* for the system.

Let us consider what forms a solution might take.

Inconsistent system (no solution)

Clearly the system

$$x_1 = 1$$
$$x_1 = 2$$

has no solution. Such a system is called *inconsistent*; if a solution does exist the system is said to be *consistent*.

Another example of an inconsistent system is

$$x_1 + x_2 = 1,$$
$$x_1 + x_2 = 0.$$

Geometrically, these equations represent parallel lines in $\mathbb{R}^2$. Solving the system corresponds to finding points of intersection: $x_1 = c_1, x_2 = c_2$ is a solution precisely when (c_1, c_2) lies on both lines. No solution exists in this case, of course, because parallel lines do not meet (Fig. 3.2).

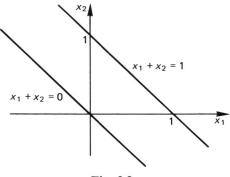

Fig. 3.2

Further examples of inconsistent systems, this time involving three unknowns, are

$$\begin{array}{ll} x_1 + x_2 + x_3 = 1 & \quad \text{and} \quad x_1 + x_2 + x_3 = 1 \\ x_1 + x_2 + x_3 = 0 & \quad \phantom{\text{and}} \quad 2x_1 + x_2 + x_3 = 3 \\ & \quad \phantom{\text{and}} \quad 3x_1 + x_2 + x_3 = 2. \end{array}$$

The first of these systems represents two parallel planes, which clearly do not intersect (Fig. 3.3(a)). The second represents a system of three planes, any

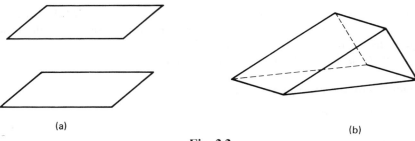

(a)　　　　　　　　　　　　　　　　(b)

Fig. 3.3

two of which intersect in a straight line, but such that *all three* do not intersect; that is, no point lies on all of the planes (Fig. 3.3(b)).

Unique solution

The system

$$x_1 = 1$$

clearly has a unique solution. Similarly, the system

$$x_1 + x_2 = 2$$
$$x_1 - x_2 = 0$$

has the unique solution $x_1 = x_2 = 1$. The geometric picture (Fig. 3.4) is of two lines intersecting in the point $(1, 1)$.

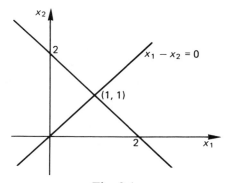

Fig. 3.4

The system

$$x_1 + x_2 + x_3 = 3$$
$$2x_1 + x_2 - x_3 = 2$$
$$-x_1 + 3x_2 + 2x_3 = 4$$

has the unique solution $x_1 = x_2 = x_3 = 1$. This corresponds to three planes intersecting in the single point $(1, 1, 1)$, such in Fig. 3.5.

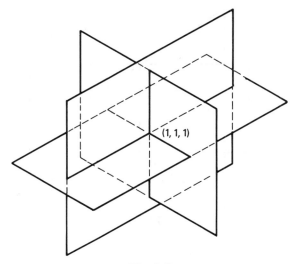

Fig. 3.5

Infinitely many solutions

The third possibility is that there may be infinitely many solutions. For example, given the system

$$x_1 + x_2 = 1,$$

we have $x_1 = 1 - x_2$, and choosing any value for x_2 gives a corresponding value for x_1.

Also, consider the system

$$x_1 + x_2 + x_3 = 1$$
$$2x_1 + x_2 + x_3 = 1$$
$$3x_1 + x_2 + x_3 = 1.$$

Subtracting the first equation from the second shows that $x_1 = 0$. Substituting into any of the equations then yields $x_2 = 1 - x_3$. So again we can choose any value for x_3 and compute a corresponding value for x_2. The geometrical interpretation of this system (Fig. 3.6) is of three planes intersecting in the line $x_1 = 0, x_2 + x_3 = 1$.

Summary

A system of linear equations may thus:

(a) be inconsistent (have no solution),

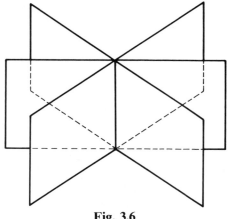

Fig. 3.6

(b) have a unique solution, or
(c) have infinitely many solutions.

EXERCISES 3.1

Interpret each of the following systems of equations geometrically in $\mathbb{R}^2$, and decide whether each of them is consistent or inconsistent.

1. $x_1 = 1$
    $x_1 + x_2 = 0.$

2. $x_1 + x_2 = 1$
    $x_1 + 2x_2 = 1.$

3. $x_1 + x_2 = 1$
    $2x_1 + 2x_2 = 3.$

4. $x_1 + x_2 = 1$
    $x_1 - x_2 = 1$
    $2x_1 + 3x_2 = 5.$

5. $x_1 + x_2 = 1$
    $x_1 - x_2 = 1$
    $2x_1 + 3x_2 = 2.$

3.2. EQUIVALENT SYSTEMS

Two systems of linear equations in $x_1, \ldots, x_n$ are said to be *equivalent* if they have exactly the same set of solutions. Given a system, we obtain an equivalent system by performing any of the following *elementary operations*:

Type 1 Multiplication of an equation by a non-zero scalar (that is, real or complex number).

Type 2 Addition of a multiple of one equation to another equation.

Type 3 Interchange of two equations.

Example 3.2.1

Solve the system

$$x_1 + 2x_2 - 3x_3 = 0, \tag{1}$$
$$2x_1 - 2x_2 - x_3 = -1, \tag{2}$$
$$-3x_1 + 5x_2 + x_3 = 3. \tag{3}$$

Solution First we eliminate x_1 from equations (2) and (3):

$$(2) - 2 \times (1): \qquad -6x_2 + 5x_3 = -1; \tag{4}$$
$$(3) + 3 \times (1): \qquad 11x_2 - 8x_2 = 3. \tag{5}$$

We now have the equivalent system (1), (4), (5). Next eliminate x_2 from equation (5):

$$(5) + \tfrac{11}{6} \times (4): \qquad \tfrac{7}{6} x_3 = \tfrac{7}{6}. \tag{6}$$

The system (1), (4), (6) is also equivalent to the original system. Now it is clear from (6) that $x_3 = 1$, and substituting back into (4) and then into (1) gives $x_1 = x_2 = 1$. The original system, therefore, has the unique solution

$$x_1 = x_2 = x_3 = 1.$$

In the above example we could omit the unknowns and simply work with the coefficients and the numbers on the right-hand sides of the equations, written in the form of a matrix. We could then present the operations on the system as follows:

$$\begin{bmatrix} 1 & 2 & -3 & \vdots & 0 \\ 2 & -2 & -1 & \vdots & -1 \\ -3 & 5 & 1 & \vdots & 3 \end{bmatrix}$$
$$\downarrow$$
$$\begin{bmatrix} 1 & 2 & -3 & \vdots & 0 \\ 0 & -6 & 5 & \vdots & -1 \\ 0 & 11 & -8 & \vdots & 3 \end{bmatrix} \quad \begin{array}{l} R_2 = r_2 - 2r_1 \\ R_3 = r_3 + 3r_1 \end{array}$$
$$\downarrow$$
$$\begin{bmatrix} 1 & 2 & -3 & \vdots & 0 \\ 0 & -6 & 5 & \vdots & -1 \\ 0 & 0 & \tfrac{7}{6} & \vdots & \tfrac{7}{6} \end{bmatrix} \quad R_3 = r_3 + \tfrac{11}{6} r_2.$$

Here we have separated the coefficients from the right-hand sides of the equations by means of a broken line, in order to avoid any possible confusion. The equation $R_2 = r_2 - 2r_1$ indicates that the new row 2 is the old row 2 minus twice the old row 1: this gives a shorthand way of indicating which elementary operations have been performed.

The system solved in example 3.2.1 had a unique solution. We look at what happens when we apply the above procedure to an inconsistent system or to a system with infinitely many solutions.

Examples 3.2.2

Solve (if possible) the systems

1. $\quad x_1 + x_2 + x_3 = 1$
   $\quad 2x_1 + 3x_2 + 3x_3 = 3$
   $\quad x_1 + 2x_2 + 2x_3 = 5,$

2. $\quad x_1 + x_2 + x_3 = 1$
   $\quad 2x_1 + x_2 + x_3 = 1$
   $\quad 3x_1 + x_2 + x_3 = 1.$

Solutions

1.
$$\begin{bmatrix} 1 & 1 & 1 & \vdots & 1 \\ 2 & 3 & 3 & \vdots & 3 \\ 1 & 2 & 2 & \vdots & 5 \end{bmatrix}$$
$$\downarrow$$
$$\begin{bmatrix} 1 & 1 & 1 & \vdots & 1 \\ 0 & 1 & 1 & \vdots & 1 \\ 0 & 1 & 1 & \vdots & 4 \end{bmatrix} \begin{array}{l} R_2 = r_2 - 2r_1 \\ R_3 = r_3 - r_1 \end{array}$$
$$\downarrow$$
$$\begin{bmatrix} 1 & 1 & 1 & \vdots & 1 \\ 0 & 1 & 1 & \vdots & 1 \\ 0 & 0 & 0 & \vdots & 3 \end{bmatrix} \; R_3 = r_3 - r_2.$$

The final line of this last matrix shows that the system is inconsistent: clearly there is no solution to the equation $0x_1 + 0x_2 + 0x_3 = 3$. In fact the inconsistency was apparent to the previous stage, as the system

$$x_2 + x_3 = 1,$$
$$x_2 + x_3 = 4$$

could hardly be consistent.

2. $$\begin{bmatrix} 1 & 1 & 1 & \vdots & 1 \\ 2 & 1 & 1 & \vdots & 1 \\ 3 & 1 & 1 & \vdots & 1 \end{bmatrix}$$

$$\downarrow$$

$$\begin{bmatrix} 1 & 1 & 1 & \vdots & 1 \\ 0 & -1 & -1 & \vdots & -1 \\ 0 & -2 & -2 & \vdots & -2 \end{bmatrix} \begin{array}{l} R_2 = r_2 - 2r_1 \\ R_3 = r_3 - 3r_1 \end{array}$$

$$\downarrow$$

$$\begin{bmatrix} 1 & 1 & 1 & \vdots & 1 \\ 0 & -1 & -1 & \vdots & -1 \\ 0 & 0 & 0 & \vdots & 0 \end{bmatrix} \quad R_3 = r_3 - 2r_2.$$

The final row of zeros here indicates that there are infinitely many solutions. Choosing any value of x_3, we then have $-x_2 - x_3 = -1$, so that $x_2 = 1 - x_3$ and $x_1 + x_2 + x_3 = 1$, giving $x_1 = 1 - x_2 - x_3 = 0$. The solution, therefore, is $x_1 = 0, x_2 = 1 - \alpha, x_3 = \alpha$, where α is any scalar. Again, the nature of the solution could have been spotted from the second matrix, where the third row is clearly twice the second row; this means that, in effect, we have only *two* equations in *three* unknowns.

EXERCISES 3.2

1 Solve (if possible) the following systems of linear equations. In each case interpret the result geometrically in $\mathbb{R}^3$.

(a) $2x_1 + 3x_2 - x_3 = -1$
$-x_1 - 4x_2 + 5x_3 = 3$
$x_1 - 2x_2 - 3x_3 = 3.$

(b) $x_1 + 2x_2 - x_3 = 1$
$3x_1 - 5x_2 + 2x_3 = 6$
$-x_1 + 9x_2 - 4x_3 = -4.$

(c) $x_1 + x_2 + x_3 = 3$
$x_1 - x_2 - 3x_3 = 3$
$-4x_1 + x_2 - 2x_3 = 4.$

(d) $x_1 + 3x_2 + x_3 = 0$
$2x_1 - x_2 - x_3 = 1$
$x_1 - 4x_2 - 2x_3 = 2.$

(e) $x_1 - x_2 - x_3 = 5$
$x_1 + x_2 + x_3 = 1$
$3x_1 - x_2 - x_3 = 11.$

(f) $3x_1 - 2x_2 + x_3 = 4$
$2x_1 + x_2 + 2x_3 = 2$
$-x_1 - 3x_2 + 3x_3 = 16.$

2 Solve (if possible) the following systems of linear equations:

(a) $x_1 + 2x_2 + 3x_3 - x_4 = 0$
$x_1 - x_2 + 4x_3 + 2x_4 = 1$
$2x_1 - 3x_2 + x_3 + x_4 = -1$
$2x_1 - x_2 + 4x_3 + x_4 = 1.$

(b) $x_1 + x_2 - x_3 - x_4 = -1$
$3x_1 + 4x_2 - x_3 - 2x_4 = 3$
$x_1 + 2x_2 + x_3 = 5.$

3.3 ELEMENTARY ROW OPERATIONS ON MATRICES

In this section and the next we will look a little more closely at the manipulations which were utilised in the previous section. As we saw, systems of linear equations can be solved by reducing them to simpler systems by using a standard set of operations. It was easier to work with just the matrix of numbers, and so we define the following operations on a matrix to be *elementary row operations*, or e.r.o.s for short.

Type 1   Multiplication of a row by a non-zero real or complex number.

Type 2   Addition to a row of a multiple of a different row.

Type 3   Interchange of two rows.

Elementary column operations, or e.c.o.s, can be defined similarly.

Examples 3.3.1

1. Performing the type 1 e.r.o. of multiplying row 2 by $\frac{1}{2}$ on the matrix $\begin{bmatrix} 1 & -1 \\ 2 & 4 \end{bmatrix}$ produces the matrix $\begin{bmatrix} 1 & -1 \\ 1 & 2 \end{bmatrix}$ $(R_2 = \frac{1}{2}r_1)$.

2. Performing the type 2 e.c.o. of subtracting twice column 2 from column 1 on the matrix $\begin{bmatrix} 1 & -1 \\ -1 & 2 \end{bmatrix}$ produces the matrix $\begin{bmatrix} 3 & -1 \\ -5 & 2 \end{bmatrix}$ $(C_1 = c_1 - 2c_2)$.

We will say that the matrix B is *row-equivalent* to the matrix A if B can be obtained from A by performing a finite sequence of elementary row operations on A. We write $B \sim A$. As it is clear that performing an e.r.o. cannot alter the size of a matrix, it is necessary that A and B have the same size.

We could speak of *column equivalences* similarly, but we shall usually restrict our attention to row equivalence, as nothing is lost by doing this.

Let us list some properties of row equivalence.

(a) Every matrix is row-equivalent to itself. This is not a very profound observation: we simply perform the e.r.o. of type 1 of multiplying any row by 1.

(b) If B is row-equivalent to A, then A is row-equivalent to B. In order to establish this it is sufficient to show that each e.r.o. has an inverse which is also an e.r.o. For, if B is obtained by performing a particular sequence of e.r.o.s to A, then we have only to perform the sequence of inverse e.r.o.s in the reverse order to B in order to obtain A.

$$A \to \ldots\ldots\ldots \to B$$
$$A \leftarrow \ldots\ldots\ldots \leftarrow B.$$

Now, multiplication of row r by α has as its inverse multiplication of row r by $1/\alpha$; addition to row r of α times row s has as its inverse subtraction from row r of α times row s; the interchange of two rows is self-inverse.

(c) If B is row-equivalent to A, and C is row-equivalent to B, then C is row-equivalent to A. For, if there is a sequence of e.r.o.s which applied to A produces B, and a sequence which applied to B produces C, then applying the first sequence followed by the second to A will result in C.

Example 3.3.2

The examples above show that the matrices $\begin{bmatrix} 1 & -1 \\ 2 & 4 \end{bmatrix}$ and $\begin{bmatrix} 1 & -1 \\ 1 & 2 \end{bmatrix}$ are row-equivalent, and that the matrices $\begin{bmatrix} 1 & -1 \\ -1 & 2 \end{bmatrix}$ and $\begin{bmatrix} 3 & -1 \\ -5 & 2 \end{bmatrix}$ are column-equivalent.

Without properties (b) and (c) above the procedures used in section 3.2 would be very suspect, of course!

EXERCISES 3.3

Look at the examples and your solutions to exercises 3.2 and describe each operation performed on the corresponding matrix of coefficients as type 1, 2 or 3.

3.4 THE REDUCED ECHELON FORM FOR A MATRIX

The basic idea of section 2 was to apply elementary operations to a system of linear equations in order to produce a simpler system. We have seen that this is equivalent to applying elementary row operations to a corresponding matrix in order to 'simplify' it. But just *how* 'simple' can we make a general matrix by means of these operations?

A matrix is said to be in *reduced echelon* form (or *Hermite normal* form) if it satisfies the following:

E1 All of the zero rows are at the bottom.
E2 The first non-zero entry in each row is 1.
E3 If the first non-zero entry in row i appears in column j, then every other element in column j is zero.
E4 The first non-zero entry in row i is strictly to the left of the first non-zero entry in row $i+1$.

This may seem quite complicated on first reading, and the information is perhaps easier to understand visually. These conditions are really saying that the non-zero elements of the matrix appear in the following 'staircase' pattern:

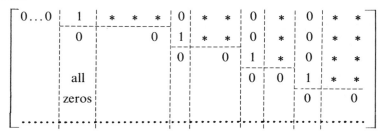

We will show now that every matrix can be reduced to one in reduced echelon form.

THEOREM 3.4.1 Every $m \times n$ matrix is row-equivalent to an $m \times n$ matrix in reduced echelon form.

Proof We consider the zero matrix to be a trivial case of reduced echelon form, so suppose that $A \neq O$. Let column k_1 be the first column containing a non-zero element. By a row interchange, if necessary, we can assume that the $(1, k_1)$-element, a_{1k_1}, is non-zero. Divide row 1 by a_{1k_1}. If $a_{jk_1} \neq 0$ for any $j \neq 1$ then subtract a_{jk_1} times row 1 from row j. This produces the following form:

$$\begin{bmatrix} 0 \ldots 0 & 1 & b_{1(k_1+1)} \ldots b_{1n} \\ & 0 & \\ & \vdots & \\ 0 \ldots 0 & 0 & b_{m(k_1+1)} \ldots b_{mn} \end{bmatrix}$$

Now let k_2 be the first column containing a non-zero element in a row other than row 1. Again, by interchanging row 2 with row i ($i \geq 2$) we can assume that $b_{2k_2} \neq 0$. Divide row 2 by b_{2k_2}. If $b_{jk_2} \neq 0$ for any $j \neq 2$ then subtract b_{jk_2} times row 2 from row j. The matrix will then have the following form:

$$\begin{bmatrix} 0\ldots 0 & 1 & *\ldots * & 0 & c_{1(k_2+1)}\ldots c_{1n} \\ 0\ldots 0 & 0 & 0\ldots 0 & 1 & \ldots\ldots\ldots\ldots \\ & & & 0 & \\ & & & \vdots & \\ 0\ldots 0 & 0 & 0\ldots 0 & 0 & c_{m(k_2+1)}\ldots c_{mn} \end{bmatrix}$$

Next look for the first column containing a non-zero element in a row other than row 1 or 2, and bring it into row 3 by a row interchange (if necessary). Continuing in the same way produces the desired form.

This proof is no thing of beauty, but this is due largely to the notation. It is, in fact, very useful because it tells us exactly what to do in any particular example. If you have found the proof indigestible, then working through an example may well help.

Example 3.4.1

Find the reduced echelon form of the matrix

$$\begin{bmatrix} 0 & 0 & 0 & 0 & 0 & 0 & 1 & -1 \\ 0 & 0 & 0 & 0 & 2 & 4 & 4 & 2 \\ 0 & 1 & 2 & 3 & 3 & 5 & 7 & 6 \\ 0 & 1 & 2 & 3 & 4 & 7 & 9 & 7 \end{bmatrix}$$

Solution

$$\begin{bmatrix} 0 & 0 & 0 & 0 & 0 & 0 & 1 & -1 \\ 0 & 0 & 0 & 0 & 2 & 4 & 4 & 2 \\ 0 & 1 & 2 & 3 & 3 & 5 & 7 & 6 \\ 0 & 1 & 2 & 3 & 4 & 7 & 9 & 7 \end{bmatrix}$$
$\downarrow$
$$\begin{bmatrix} 0 & 1 & 2 & 3 & 3 & 5 & 7 & 6 \\ 0 & 0 & 0 & 0 & 2 & 4 & 4 & 2 \\ 0 & 0 & 0 & 0 & 0 & 0 & 1 & -1 \\ 0 & 1 & 2 & 3 & 4 & 7 & 9 & 7 \end{bmatrix} \begin{matrix} R_1 = r_3 \\ \\ R_3 = r_1 \\ \end{matrix}$$
$\downarrow$
$$\begin{bmatrix} 0 & 1 & 2 & 3 & 3 & 5 & 7 & 6 \\ 0 & 0 & 0 & 0 & 2 & 4 & 4 & 2 \\ 0 & 0 & 0 & 0 & 0 & 0 & 1 & -1 \\ 0 & 0 & 0 & 0 & 1 & 2 & 2 & 1 \end{bmatrix} \begin{matrix} \\ \\ \\ R_4 = r_4 - r_1 \end{matrix}$$
$\downarrow$

$$\begin{bmatrix} 0 & 1 & 2 & 3 & 3 & 5 & 7 & 6 \\ 0 & 0 & 0 & 0 & 1 & 2 & 2 & 1 \\ 0 & 0 & 0 & 0 & 0 & 0 & 1 & -1 \\ 0 & 0 & 0 & 0 & 1 & 2 & 2 & 1 \end{bmatrix} \quad R_2 = \tfrac{1}{2}r_2$$

$$\downarrow$$

$$\begin{bmatrix} 0 & 1 & 2 & 3 & 0 & -1 & 1 & 3 \\ 0 & 0 & 0 & 0 & 1 & 2 & 2 & 1 \\ 0 & 0 & 0 & 0 & 0 & 0 & 1 & -1 \\ 0 & 0 & 0 & 0 & 0 & 0 & 0 & 0 \end{bmatrix} \quad \begin{aligned} R_1 &= r_1 - 3r_2 \\ \\ \\ R_4 &= r_4 - r_2 \end{aligned}$$

$$\downarrow$$

$$\begin{bmatrix} 0 & 1 & 2 & 3 & 0 & -1 & 0 & 4 \\ 0 & 0 & 0 & 0 & 1 & 2 & 0 & 3 \\ 0 & 0 & 0 & 0 & 0 & 0 & 1 & -1 \\ 0 & 0 & 0 & 0 & 0 & 0 & 0 & 0 \end{bmatrix} \quad \begin{aligned} R_1 &= r_1 - r_3 \\ R_2 &= r_2 - 2r_3. \end{aligned}$$

We can apply this method to the matrix associated with a system of linear equations and thereby produce a simpler equivalent system. This, however, involves more work than is strictly necessary. It suffices to reduce to a matrix satisfying E1 and E4: such a matrix is said to be in *echelon form*, and it follows from the above theorem that any matrix can be row-reduced to one in echelon form.

Example 3.4.2

The matrices
$$\begin{bmatrix} 1 & 1 & 1 & 1 \\ 0 & 1 & 1 & 1 \\ 0 & 0 & 0 & 3 \end{bmatrix}, \begin{bmatrix} 1 & 1 & 1 & 1 \\ 0 & -1 & -1 & -1 \\ 0 & 0 & 0 & 0 \end{bmatrix},$$

$$\begin{bmatrix} 1 & 2 & -3 & 0 \\ 0 & -6 & 5 & -1 \\ 0 & 0 & \tfrac{7}{6} & \tfrac{7}{6} \end{bmatrix}$$

encountered in the previous section are in echelon form, but none of them is in reduced echelon form.

A matrix has infinitely many different echelon forms, but its reduced echelon form is unique. As a consequence of this fact, which we hope the reader will take on trust, we will often refer to *the* reduced echelon form of a given matrix.

EXERCISES 3.4

1 Row-reduce the following matrices to reduced echelon form:

(a) $\begin{bmatrix} 2 & 4 & -2 & 0 & 12 & 0 \\ 0 & -1 & -8 & -3 & -9 & 2 \\ -2 & -4 & 2 & 3 & 0 & -3 \\ 2 & 3 & -10 & 0 & 15 & 0 \end{bmatrix}$;

(b) $\begin{bmatrix} 1 & 6 & 2 \\ 1 & -3 & -1 \\ 2 & 9 & 1 \end{bmatrix}$; (c) $\begin{bmatrix} 5 & 6 & 8 & -1 \\ 4 & 3 & 0 & 0 \\ 10 & 12 & 16 & -2 \\ 1 & 2 & 0 & 0 \end{bmatrix}$.

2 Describe the general appearance of a matrix in echelon form.

3 What does a square matrix in echelon form look like?

4 Show that the reduced echelon form of an $n \times n$ matrix either is I_n or else contains at least one row of zeros.

3.5 ELEMENTARY MATRICES

The result of applying an e.r.o. to an identity matrix, I_n, is called an *elementary matrix (of size n)*. We thus have the following types of elementary matrix:

Type 1 $\quad E_r(\alpha) = \begin{bmatrix} 1 \dots 0 \dots 0 \\ \dots\dots\dots \\ 0 \dots \alpha \dots 0 \\ \dots\dots\dots \\ 0 \dots 0 \dots 1 \end{bmatrix} \leftarrow \text{row } r$

$\qquad\qquad\qquad\qquad\quad\uparrow$
$\qquad\qquad\qquad\quad\text{column } r$

Type 2 $\quad E_{rs}(\alpha) = \begin{bmatrix} 1 \dots 0 \dots 0 \dots 0 \\ \dots\dots\dots\dots \\ 0 \dots 1 \dots \alpha \dots 0 \\ \dots\dots\dots\dots \\ 0 \dots 0 \dots 0 \dots 1 \end{bmatrix} \leftarrow \text{row } r$

$\qquad\qquad\qquad\qquad\quad\uparrow\quad\uparrow$
$\qquad\qquad\qquad\text{column } r\;\;\text{column } s$

Type 3 $E_{rs} = \begin{bmatrix} 1...0...0...0 \\ .... \\ 0...0...1...0 \\ .... \\ 0...1...0...0 \\ .... \\ 0...0...0...1 \end{bmatrix}$ $\begin{matrix} \\ \\ \leftarrow \text{row } r \\ \\ \leftarrow \text{row } s \\ \\ \\ \end{matrix}$

$\phantom{Type 3 \quad E_{rs} = }\ \ \uparrow\ \ \uparrow$
$\phantom{Type 3 \quad E_{rs} = }\ $column r column s

Example 3.5.1

Let $n = 3$. Then

$$E_2(3) = \begin{bmatrix} 1 & 0 & 0 \\ 0 & 3 & 0 \\ 0 & 0 & 1 \end{bmatrix}, \ E_{13}(5) = \begin{bmatrix} 1 & 0 & 5 \\ 0 & 1 & 0 \\ 0 & 0 & 1 \end{bmatrix}, \ E_{13} = \begin{bmatrix} 0 & 0 & 1 \\ 0 & 1 & 0 \\ 1 & 0 & 0 \end{bmatrix}.$$

Now let us consider the effects of multiplying a matrix A by an elementary matrix. We shall refer to multiplying A *on the left* by a matrix as *pre-multiplication* by that matrix; likewise, *post-multiplication* refers to multiplication *on the right*. Let us look first of all at pre-multiplication of $A = [a_{ij}]_{m \times n}$ by an elementary matrix of size m.

Type 1 $\quad E_r(\alpha)A = \begin{bmatrix} 1...0...0 \\ .... \\ 0...\alpha...0 \\ .... \\ 0...0...1 \end{bmatrix} \begin{bmatrix} a_{11}...a_{1r}...a_{1n} \\ .... \\ a_{r1}...a_{rr}...a_{rn} \\ .... \\ a_{m1}...a_{mr}...a_{mn} \end{bmatrix}$

$$= \begin{bmatrix} a_{11}...a_{1r}...a_{1n} \\ .... \\ \alpha a_{r1}...\alpha a_{rr}...\alpha a_{rn} \\ .... \\ a_{m1}...a_{mr}...a_{mn} \end{bmatrix}.$$

The result is the same as that of performing the corresponding e.r.o. of type 1 to A.

//
Now try evaluating $E_{rs}(\alpha)A$ and $E_{rs}A$. You should find that in these cases too pre-multiplication by an elementary matrix has the same effect as performing the corresponding row operation to A.

Example 3.5.2

Let $A = \begin{bmatrix} 1 & 2 & 3 \\ 4 & 5 & 6 \\ 7 & 8 & 9 \end{bmatrix}$. Then

$$E_2(-1)A = \begin{bmatrix} 1 & 0 & 0 \\ 0 & -1 & 0 \\ 0 & 0 & 1 \end{bmatrix} \begin{bmatrix} 1 & 2 & 3 \\ 4 & 5 & 6 \\ 7 & 8 & 9 \end{bmatrix} = \begin{bmatrix} 1 & 2 & 3 \\ -4 & -5 & -6 \\ 7 & 8 & 9 \end{bmatrix}.$$

$$E_{31}(1)A = \begin{bmatrix} 1 & 0 & 0 \\ 0 & 1 & 0 \\ 1 & 0 & 1 \end{bmatrix} \begin{bmatrix} 1 & 2 & 3 \\ 4 & 5 & 6 \\ 7 & 8 & 9 \end{bmatrix} = \begin{bmatrix} 1 & 2 & 3 \\ 4 & 5 & 6 \\ 8 & 10 & 12 \end{bmatrix}.$$

$$E_{13}A = \begin{bmatrix} 0 & 0 & 1 \\ 0 & 1 & 0 \\ 1 & 0 & 0 \end{bmatrix} \begin{bmatrix} 1 & 2 & 3 \\ 4 & 5 & 6 \\ 7 & 8 & 9 \end{bmatrix} = \begin{bmatrix} 7 & 8 & 9 \\ 4 & 5 & 6 \\ 1 & 2 & 3 \end{bmatrix}.$$

Similarly, it is straightforward to check that post-multiplication by an elementary matrix is equivalent to performing the corresponding e.c.o. to A.

The fact that e.r.o.s have inverses that are also e.r.o.s corresponds to the following result about elementary matrices.

THEOREM 3.5.1 Elementary matrices are invertible, and their inverses are also elementary matrices of the same type.

Proof Easy matrix multiplications are sufficient to check that
$$E_r(\alpha)E_r(\alpha^{-1}) = E_r(\alpha^{-1})E_r(\alpha) = I,$$
$$E_{rs}(\alpha)E_{rs}(-\alpha) = E_{rs}(-\alpha)E_{rs}(\alpha) = I,$$
$$E_{rs}E_{rs} = I.$$

Alternatively, the results can be checked by considering the corresponding e.r.o.s. Hence,
$$E_r(\alpha)^{-1} = E_r(\alpha^{-1}), \qquad E_{rs}(\alpha)^{-1} = E_{rs}(-\alpha), \qquad E_{rs}^{-1} = E_{rs}.$$

COROLLARY 3.5.2 If the matrix B is row-(respectively, column-)equivalent to A then there is an invertible matrix P such that $B = PA$ (respectively, $B = AP$).

Proof Suppose that B is row-equivalent to A. Then B can be produced from A by performing a finite sequence of e.r.o.s to it. Let $E_1, \ldots, E_k$ be the corresponding elementary matrices. Then $B = (E_k \ldots E_2 E_1)A$. Put $P = E_k \ldots E_2 E_1$. Then P is invertible by Theorem 3.5.1 and the generalization of I2 (see section 2.6).

If B is column-equivalent to A then the result is proved similarly.

EXERCISES 3.5

1 Write down the following elementary matrices of size 4:

(a) $E_2(-1)$; (b) E_{24}; (c) $E_{12}(-3)$; (d) $E_4(3)$; (e) E_{21}; (f) $E_{31}(2)$.

2 Pre- and post-multiply the matrix $A = \begin{bmatrix} 1 & 0 & -1 \\ 2 & 3 & 1 \\ -1 & -1 & 2 \end{bmatrix}$ by each of the following elementary matrices of size 3:

(a) $E_{13}(4)$; (b) $E_1(6)$; (c) E_{13}; (d) $E_{23}(-5)$.

3 Check that pre- (respectively, post-) multiplication of $A = [a_{ij}]_{m \times n}$ by the elementary matrix $E_{rs}(\alpha)$ of size m has the same effect as performing an e.r.o. (respectively, e.c.o.) of type 2 on A.

4 Check that pre- (respectively, post-) multiplication of $A = [a_{ij}]_{m \times n}$ by the elementary matrix E_{rs} of size m has the same effect as performing an e.r.o. (respectively, e.c.o.) of type 3 on A.

5 Let A be the matrix in exercise 3.4.1(a) and let B be its reduced echelon form. Find an invertible matrix P such that $B = PA$. Repeat with A representing the other matrices in exercise 3.4.1.

6 Prove that if the matrix B is column-equivalent to A then there is an invertible matrix P such that $B = AP$.

3.6 FINDING THE INVERSE OF AN INVERTIBLE MATRIX

We can use the results of the previous sections to answer question (a) on p. 44; that is, we can deduce a method for inverting an invertible matrix.

Let A be a square matrix and suppose that A is row-equivalent to an identity matrix. Then A can be row-reduced to I by a sequence of e.r.o.s. Let $E_1, E_2, \ldots, E_r$ be the corresponding elementary matrices, so that

$$E_r \ldots E_2 E_1 A = I. \tag{\dagger}$$

But $E_1, E_2, \ldots, E_r$ are invertible, by Theorem 3.5.1, and multiplying both sides of (†) by $E_1^{-1} E_2^{-1} \ldots E_r^{-1}$ gives $A = E_1^{-1} E_2^{-1} \ldots E_r^{-1}$. Since A is a product of elementary matrices, it is invertible and

$$A^{-1} = (E_1^{-1} E_2^{-1} \ldots E_r^{-1})^{-1} = (E_r \ldots E_2 E_1) I.$$

We have proved the following.

THEOREM 3.6.1 If the square matrix A can be row-reduced to an identity matrix by a sequence of e.r.o.s then A is invertible and the inverse of A is found by applying the same sequence of e.r.o.s to I.

Example 3.6

We now illustrate the above theorem by applying it to the matrix
$\begin{bmatrix} 1 & 0 & -1 \\ 1 & -1 & 1 \\ 0 & 0 & -1 \end{bmatrix}$. If A is row-equivalent to I, then I is the reduced echelon form of A. (Look back to section 3.4 to convince yourself of this.) We can, therefore, use the procedure described in Theorem 3.4.1 in order to row-reduce A to I. In carrying out this reduction we apply each e.r.o. simultaneously to A and to I. In order to do this it is convenient to write down A and I side by side (but separated by a dotted line) as follows:

$$\begin{array}{c} A \\ \downarrow \end{array} \qquad \begin{array}{c} I \\ \downarrow \end{array}$$

$$\left[\begin{array}{ccc|ccc} 1 & 0 & -1 & 1 & 0 & 0 \\ 1 & -1 & 1 & 0 & 1 & 0 \\ 0 & 0 & -1 & 0 & 0 & 1 \end{array} \right]$$

$$\downarrow$$

$$\left[\begin{array}{ccc|ccc} 1 & 0 & -1 & 1 & 0 & 0 \\ 0 & -1 & 2 & -1 & 1 & 0 \\ 0 & 0 & 1 & 0 & 0 & -1 \end{array} \right] \begin{array}{l} R_2 = r_2 - r_1 \\ R_3 = -r_3 \end{array}$$

$$\downarrow$$

$$\left[\begin{array}{ccc|ccc} 1 & 0 & -1 & 1 & 0 & 0 \\ 0 & 1 & -2 & 1 & -1 & 0 \\ 0 & 0 & 1 & 0 & 0 & -1 \end{array} \right] R_2 = -r_2$$

$$\begin{bmatrix} 1 & 0 & 0 & \vdots & 1 & 0 & -1 \\ 0 & 1 & 0 & \vdots & 1 & -1 & -2 \\ 0 & 0 & 1 & \vdots & 0 & 0 & -1 \end{bmatrix} \begin{array}{l} R_1 = r_1 + r_3 \\ R_2 = r_2 + 2r_3 \end{array}$$

$\uparrow$ I $\quad\quad$ $\uparrow$ This is A^{-1}

The inverse matrix is, therefore, $\begin{bmatrix} 1 & 0 & -1 \\ 1 & -1 & -2 \\ 0 & 0 & -1 \end{bmatrix}$.

Note: It is very easy to make a simple arithmetical slip in all of this working, so that it makes sense to check that $AA^{-1} = A^{-1}A = I$.

EXERCISES 3.6

1 Find the inverses of the following matrices:

(a) $\begin{bmatrix} 0 & 1 & 2 \\ 2 & 1 & 3 \\ 1 & 2 & 4 \end{bmatrix}$; (b) $\begin{bmatrix} 5 & 2 & 2 \\ 2 & 2 & -4 \\ 2 & -4 & 2 \end{bmatrix}$; (c) $\begin{bmatrix} 5 & 1 & -1 \\ 2 & 5 & 1 \\ 1 & -2 & 2 \end{bmatrix}$.

2 Let A be an $n \times n$ matrix and let B be its reduced echelon form. Prove that
(a) $B = I_n$ if and only if A is invertible;
(b) B has at least one row of zeros if and only if A is not invertible.
(Look back at exercise 3.4.4.)

3 Use the result of question 2 to show that the following matrices are not invertible:

(a) $\begin{bmatrix} 1 & -1 & 2 \\ 1 & 3 & 1 \\ 3 & 1 & 5 \end{bmatrix}$; (b) $\begin{bmatrix} 1 & 0 & 2 \\ -1 & 3 & 1 \\ 1 & -2 & 0 \end{bmatrix}$.

SOLUTIONS AND HINTS FOR EXERCISES

Exercises 3.1

1 This represents two lines intersecting in the point $(1, -1)$; it is consistent.
2 This represents two lines intersecting in the point $(1, 0)$; it is consistent.
3 This represents two parallel lines; it is inconsistent.
4 This represents three lines forming the sides of a triangle; it is inconsistent.
5 This represents three lines intersecting in the point $(1, 0)$; it is consistent.

ROW REDUCTION

Exercises 3.2

1. (a) This system has the unique solution $x_1 = 1$, $x_2 = -1$, $x_3 = 0$ and so represents three planes intersecting in a single point.
   (b) The solution is $x_1 = \frac{1}{11}x_3 + \frac{17}{11}$, $x_2 = \frac{5}{11}x_3 - \frac{3}{11}$, where x_3 is any real number. The system represents three planes intersecting on a line.
   (c) The solution is $x_1 = 1$, $x_2 = 4$, $x_3 = -2$. The system represents three planes intersecting in a single point.
   (d) This system is inconsistent; it represents three planes, any two of which intersect in a line, but all three of which do not intersect.
   (e) The solution is $x_1 = 3$, $x_2 = -x_3 - 2$, where x_3 is any real number. The system represents three planes intersecting in a line.
   (f) The solution is $x_1 = -1$, $x_2 = -2$, $x_3 = 3$. The system represents three planes intersecting in a single point.

2. (a) This system is inconsistent.
   (b) The solution is $x_1 = 3x_3 + 2x_4 - 7$, $x_2 = 6 - 2x_3 - x_4$, where x_3, x_4 are any real numbers.

Exercises 3.4

1. The reduced echelon forms are

   (a) $\begin{bmatrix} 1 & 0 & -17 & 0 & 12 & 0 \\ 0 & 1 & 8 & 0 & -3 & 0 \\ 0 & 0 & 0 & 1 & 4 & 0 \\ 0 & 0 & 0 & 0 & 0 & 1 \end{bmatrix}$; (b) $\begin{bmatrix} 1 & 0 & 0 \\ 0 & 1 & 0 \\ 0 & 0 & 1 \end{bmatrix}$;

   (c) $\begin{bmatrix} 1 & 0 & 0 & 0 \\ 0 & 1 & 0 & 0 \\ 0 & 0 & 1 & -\frac{1}{8} \\ 0 & 0 & 0 & 0 \end{bmatrix}$.

2. A matrix in echelon form looks like

   $$\begin{bmatrix} 0\ldots0 & * & \cdots & & & & \\ & & 0 & \ldots & 0 & * & \cdots \\ & \text{all} & & & 0 & . & 0 & * & \cdots \\ & \text{zeros} & & & & & 0 & * & \cdots \end{bmatrix}$$

   where * indicates an element which *must* be non-zero.

3. A square matrix $A = [a_{ij}]_{n \times n}$ in echelon form is upper triangular (that is, $a_{ij} = 0$ if $i > j$).

4. Let $A = [a_{ij}]_{n \times n}$ be in reduced echelon form. It is clear from E4 that $a_{ij} = 0$ if $i > j$. Suppose that $a_{ii} = 0$ for some $1 \leq i \leq n$. If the first non-zero entry in

row i is a_{ij}, then $j > i$. But now E4 implies that row $i + (n-j) + 1 (\leqslant n)$ is a row of zeros.

If $a_{ii} \neq 0$ for every $1 \leqslant i \leqslant n$, then E2 implies that $a_{ii} = 1$ for every $1 \leqslant i \leqslant n$. Furthermore, it follows from E3 that $a_{ij} = 0$ if $i \neq j$. In this case, therefore, $A = I_n$.

Exercises 3.5

1 (a) $\begin{bmatrix} 1 & 0 & 0 & 0 \\ 0 & -1 & 0 & 0 \\ 0 & 0 & 1 & 0 \\ 0 & 0 & 0 & 1 \end{bmatrix}$; (b) $\begin{bmatrix} 1 & 0 & 0 & 0 \\ 0 & 0 & 0 & 1 \\ 0 & 0 & 1 & 0 \\ 0 & 1 & 0 & 0 \end{bmatrix}$;

(c) $\begin{bmatrix} 1 & -3 & 0 & 0 \\ 0 & 1 & 0 & 0 \\ 0 & 0 & 1 & 0 \\ 0 & 0 & 0 & 1 \end{bmatrix}$; (d) $\begin{bmatrix} 1 & 0 & 0 & 0 \\ 0 & 1 & 0 & 0 \\ 0 & 0 & 1 & 0 \\ 0 & 0 & 0 & 3 \end{bmatrix}$;

(e) $\begin{bmatrix} 0 & 1 & 0 & 0 \\ 1 & 0 & 0 & 0 \\ 0 & 0 & 1 & 0 \\ 0 & 0 & 0 & 1 \end{bmatrix}$; (f) $\begin{bmatrix} 1 & 0 & 0 & 0 \\ 0 & 1 & 0 & 0 \\ 2 & 0 & 1 & 0 \\ 0 & 0 & 0 & 1 \end{bmatrix}$.

2 (a) $\begin{bmatrix} -3 & -4 & 7 \\ 2 & 3 & 1 \\ -1 & -1 & 2 \end{bmatrix}$; (b) $\begin{bmatrix} 6 & 0 & -6 \\ 2 & 3 & 1 \\ -1 & -1 & 2 \end{bmatrix}$;

(c) $\begin{bmatrix} -1 & -1 & 2 \\ 2 & 3 & 1 \\ 1 & 0 & -1 \end{bmatrix}$; (d) $\begin{bmatrix} 1 & 0 & -1 \\ 7 & 8 & -9 \\ -1 & -1 & 2 \end{bmatrix}$.

3, 4 These are straightforward.

5 (a) $\begin{bmatrix} -\frac{3}{2} & 0 & 0 & 2 \\ 1 & 0 & 0 & -1 \\ -\frac{5}{3} & -1 & -\frac{2}{3} & 1 \\ -2 & -1 & -1 & 1 \end{bmatrix}$; (b) $(\frac{1}{6}) \begin{bmatrix} 2 & 4 & 0 \\ -1 & -1 & 1 \\ 5 & 1 & -3 \end{bmatrix}$;

(c) $(\frac{1}{40}) \begin{bmatrix} 0 & 16 & 0 & -24 \\ 0 & -8 & 0 & 32 \\ 5 & -4 & 0 & -9 \\ -80 & 0 & 40 & 0 \end{bmatrix}$.

6 Suppose that B is column-equivalent to A. Then B can be produced from A by performing a finite sequence of e.c.o.s. to it. Let $E_1,\ldots,E_k$ be the corresponding elementary matrices. Then $B = A(E_1 \ldots E_k)$. Put $P = E_1 \ldots E_k$. Then P is invertible by Theorem 3.5.1 and the generalisation of I2.

Exercises 3.6

1 (a) $\begin{bmatrix} -2 & 0 & 1 \\ -5 & -2 & 4 \\ 3 & 1 & -2 \end{bmatrix}$; (b) $(\tfrac{1}{18})\begin{bmatrix} 2 & 2 & 2 \\ 2 & -1 & -4 \\ 2 & -4 & -1 \end{bmatrix}$;

(c) $(\tfrac{1}{66})\begin{bmatrix} 12 & 0 & 6 \\ -3 & 11 & -7 \\ -9 & 11 & 23 \end{bmatrix}$.

2 By exercise 3.4.4, either $B = I_n$ or B contains a row of zeros. Now if $B = I_n$ then A is invertible, by Theorem 3.6.1. So let A be invertible. By Corollary 3.5.2, $B = PA$, where P is invertible, so B is invertible (property I2, section 2.6). Hence there exists B^{-1} such that $BB^{-1} = I_n$. Now B cannot have a row of zeros (since this would imply that I_n has a row of zeros, by exercise 2.4.7); thus $B = I_n$.

3 The reduced echelon forms are (a) $\begin{bmatrix} 1 & 0 & \tfrac{7}{4} \\ 0 & 1 & -\tfrac{1}{4} \\ 0 & 0 & 0 \end{bmatrix}$; (b) $\begin{bmatrix} 1 & 0 & 2 \\ 0 & 1 & 1 \\ 0 & 0 & 0 \end{bmatrix}$.

4 DETERMINANTS

4.1 THE SIGN OF A PERMUTATION

In this chapter we intend to address the second question raised in section 3.6; the idea is to associate with each square matrix a number, called its 'determinant', which will tell us whether or not the matrix is invertible. This topic may be approached in a number of ways, ranging from the informal to the elegant and sophisticated, and, as is common in this rich and varied world, none is ideal from every standpoint: difficult choices have to be made. We have chosen to reject the elegant modern approaches because of their level of abstraction, and to give a slightly dated treatment which is, nevertheless, rigorous enough for those demanding thoroughness.

The proofs to be found in this chapter are not of importance in modern linear algebra. If you find them difficult to follow, do not despair: it is sufficient to gain some technical efficiency in calculating determinants, by knowing and practising their properties, and to appreciate the theoretical roles played by the determinant.

Our approach requires first a slight digression: the definition of a determinant that we will adopt depends on the concept of a permutation and its sign. We will introduce these notions rather quickly, as they are not central to the development of linear algebra. A more complete treatment can be found in most elementary textbooks on abstract algebra, such as Whitehead.[†]

A *permutation* σ of the set $X = \{1, 2, \ldots, n\}$ is a bijective mapping from X onto itself. Such a map σ is commonly written as

$$\sigma = \begin{pmatrix} 1 & 2 & \ldots & n \\ a_1 & a_2 & \ldots & a_n \end{pmatrix}$$

where a_i is the image under σ of i. The set of all such permutations on X will be denoted by S_n. Then S_n contains $n!$ elements, since there are, in X, n choices for a_1, $(n-1)$ choices for a_2 (as the injectivity of σ requires a_2 to be different from a_1), $(n-2)$ choices for a_3, and so on.

[†] Carol Whitehead, *Guide to Abstract Algebra* (Macmillan, 1988).

DETERMINANTS

Permutations can be multiplied by means of composition of mappings, as follows:
$$\begin{pmatrix} a_1 & a_2 & \ldots & a_n \\ b_1 & b_2 & \ldots & b_n \end{pmatrix} \begin{pmatrix} 1 & 2 & \ldots & n \\ a_1 & a_2 & \ldots & a_n \end{pmatrix} = \begin{pmatrix} 1 & 2 & \ldots & n \\ b_1 & b_2 & \ldots & b_n \end{pmatrix}$$

Recall that if σ_1 and σ_2 are mappings, then $\sigma_1 \sigma_2(x) = \sigma_1(\sigma_2(x))$. Hence, if σ_1 and σ_2 are permutations, then $\sigma_1 \sigma_2$ means 'do σ_2 first, *then* do σ_1'. Since permutations are bijective mappings, they have inverses:

$$\begin{pmatrix} 1 & 2 & \ldots & n \\ a_1 & a_2 & \ldots & a_n \end{pmatrix}^{-1} = \begin{pmatrix} a_1 & a_2 & \ldots & a_n \\ 1 & 2 & \ldots & n \end{pmatrix}.$$

Examples 4.1.1

1. $\sigma = \begin{pmatrix} 1 & 2 & 3 \\ 2 & 1 & 3 \end{pmatrix}$ is the map $\sigma: \{1, 2, 3\} \to \{1, 2, 3\}$ given by $\sigma(1) = 2$, $\sigma(2) = 1$, $\sigma(3) = 3$.

2. The six elements of S_3 are $\begin{pmatrix} 1 & 2 & 3 \\ 1 & 2 & 3 \end{pmatrix}, \begin{pmatrix} 1 & 2 & 3 \\ 1 & 3 & 2 \end{pmatrix}, \begin{pmatrix} 1 & 2 & 3 \\ 2 & 1 & 3 \end{pmatrix}, \begin{pmatrix} 1 & 2 & 3 \\ 2 & 3 & 1 \end{pmatrix},$
$\begin{pmatrix} 1 & 2 & 3 \\ 3 & 1 & 2 \end{pmatrix}, \begin{pmatrix} 1 & 2 & 3 \\ 3 & 2 & 1 \end{pmatrix}.$

3. $\begin{pmatrix} 1 & 2 & 3 \\ 2 & 3 & 1 \end{pmatrix} \begin{pmatrix} 1 & 2 & 3 \\ 2 & 1 & 3 \end{pmatrix} = \begin{pmatrix} 1 & 2 & 3 \\ 3 & 2 & 1 \end{pmatrix}, \begin{pmatrix} 1 & 2 & 3 \\ 2 & 3 & 1 \end{pmatrix}^{-1} = \begin{pmatrix} 1 & 2 & 3 \\ 3 & 1 & 2 \end{pmatrix}.$

The notation $(a_1 \ a_2 \ \ldots \ a_n)$ is used to denote the permutation which maps $a_1 \mapsto a_2, a_2 \mapsto a_3, \ldots, a_{n-1} \mapsto a_n, a_n \mapsto a_1$; thus, it is another way of writing $\begin{pmatrix} a_1 & a_2 & \ldots & a_n \\ a_2 & a_3 & \ldots & a_1 \end{pmatrix}$. We will denote the *identity permutation* $\begin{pmatrix} 1 & 2 & \ldots & n \\ 1 & 2 & \ldots & n \end{pmatrix}$ by (1).

Next we introduce the polynomial $P_n(x_1, \ldots, x_n) = \prod_{i>j}(x_i - x_j)$ in the indeterminates $x_1, \ldots, x_n$, where the product is taken over all factors with $1 \leq i < j \leq n$. Given $\sigma \in S_n$ we define

$$\hat{\sigma}(P_n) = P_n(x_{\sigma(1)}, \ldots, x_{\sigma(n)}) = \prod_{i>j}(x_{\sigma(i)} - x_{\sigma(j)}).$$

Examples 4.1.2

1. $P_2(x_1, x_2) = (x_2 - x_1)$; $P_3(x_1, x_2, x_3) = (x_3 - x_2)(x_3 - x_1)(x_2 - x_1)$;
$P_4(x_1, x_2, x_3, x_4) = (x_4 - x_3)(x_4 - x_2)(x_4 - x_1)(x_3 - x_2)(x_3 - x_1)(x_2 - x_1)$.

2. If $\sigma = (12)$ then $\hat{\sigma}(P_2) = (x_{\sigma(2)} - x_{\sigma(1)}) = x_1 - x_2 = -P_2$.

3. If $\sigma = (132)$ then

$$\hat{\sigma}(P_3) = (x_{\sigma(3)} - x_{\sigma(2)})(x_{\sigma(3)} - x_{\sigma(1)})(x_{\sigma(2)} - x_{\sigma(1)})$$
$$= (x_2 - x_1)(x_2 - x_3)(x_1 - x_3) = P_3.$$

It is not difficult to see that $\hat{\sigma}(P_n)$ is always either P_n or $-P_n$. If $\hat{\sigma}(P_n) = P_n$ we call σ an *even* permutation and write $sign(\sigma) = +1$; if $\hat{\sigma}(P_n) = -P_n$ then σ is an *odd* permutation, and $sign(\sigma) = -1$. It can be shown that $sign(\sigma_1 \sigma_2) = (sign \sigma_1)(sign \sigma_2)$, and that $sign(\sigma^{-1}) = sign \sigma$.

Examples 4.1.3

1. $sign(12) = -1$, so that (12) is odd.
2. $sign(132) = +1$, so that (132) is even.
3. In the subsequent sections it will prove useful to be able to describe the elements of S_n in different ways. For example, if $\psi = (rs)$ where $1 \leq r, s \leq n$ then $S_n = \{\sigma\psi : \sigma \in S_n\}$. For, let S denote the right-hand side of this equation. Then, if $\sigma \in S_n$, clearly $\sigma\psi^{-1} \in S_n$ and so $\sigma = (\sigma\psi^{-1})\psi \in S$. Thus, $S_n \subset S$. Conversely, if $\sigma \in S_n$, then $\sigma\psi \in S_n$, and so $S \subset S_n$.

EXERCISES 4.1

1. List the elements of S_4.

2. Find the sign of each of the elements of S_3, and thereby verify that half of them are even and half odd.

3. Prove that $S_n = \{\sigma^{-1} : \sigma \in S_n\}$.

4. Let $\sigma \in S_n$. Prove that if $\sigma \neq (1)$ then there is an i lying between 1 and n such that $\sigma(i) < i$.

5. Let T_n denote the set of permutations of the set $\{2, \ldots, n\}$. Show how the definition of sign can be modified so as to apply to elements of T_n. Let $\psi \in T_n$, and define $\sigma \in S_n$ by $\sigma(i) = \begin{cases} 1 & \text{if } i = 1 \\ \psi(i) & \text{if } i > 1 \end{cases}$. Prove that $sign \psi = sign \sigma$.

4.2 THE DEFINITION OF A DETERMINANT

Let us consider when matrices of small size have inverses.

1×1 matrices

Let $A = [a_{11}]$ be a 1×1 matrix. If $a_{11} \neq 0$ then

DETERMINANTS

$$[a_{11}][1/a_{11}] = [1/a_{11}][a_{11}] = [1],$$

so that $[a_{11}]^{-1} = [1/a_{11}]$; also, $[0]$ has no inverse. Thus, A is invertible if and only if $a_{11} \neq 0$. We call the number a_{11} the *determinant*, det A, of the 1×1 matrix A.

2×2 matrices

Let $A = \begin{bmatrix} a_{11} & a_{12} \\ a_{21} & a_{22} \end{bmatrix}$. If $B = \begin{bmatrix} a_{22} & -a_{12} \\ -a_{21} & a_{11} \end{bmatrix}$ then

$$AB = \begin{bmatrix} a_{11} & a_{12} \\ a_{21} & a_{22} \end{bmatrix} \begin{bmatrix} a_{22} & -a_{12} \\ -a_{21} & a_{11} \end{bmatrix}$$

$$= \begin{bmatrix} a_{11}a_{22} - a_{12}a_{21} & 0 \\ 0 & a_{11}a_{22} - a_{12}a_{21} \end{bmatrix}.$$

Hence $A^{-1} = \dfrac{1}{a_{11}a_{22} - a_{12}a_{21}} \begin{bmatrix} a_{22} & -a_{12} \\ -a_{21} & a_{11} \end{bmatrix}$,

and so A is invertible if and only if $a_{11}a_{22} - a_{12}a_{21} \neq 0$. The number $a_{11}a_{22} - a_{12}a_{21}$ is called the *determinant* of A and is denoted by det A or by $\begin{vmatrix} a_{11} & a_{12} \\ a_{21} & a_{22} \end{vmatrix}$

3×3 matrices

In this case it can be shown that A is invertible, where

$$A = \begin{bmatrix} a_{11} & a_{12} & a_{13} \\ a_{21} & a_{22} & a_{23} \\ a_{31} & a_{32} & a_{33} \end{bmatrix},$$

if and only if

$$a_{11}a_{22}a_{33} + a_{12}a_{23}a_{31} + a_{13}a_{21}a_{32} - a_{11}a_{23}a_{32} - a_{12}a_{21}a_{33} - a_{13}a_{22}a_{31} \neq 0.$$

This number is again called the *determinant* of A; it is again denoted by det A or by the matrix elements enclosed by vertical lines. An easy way of remembering the terms in this sum and the signs in front of them is as follows:

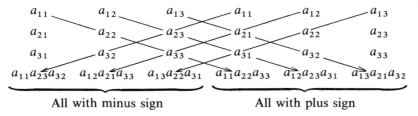

(*Note of caution:* This method of calculating the determinant does *not* generalise to matrices of other sizes.)

GUIDE TO LINEAR ALGEBRA

In order to gain some insight into how these results might generalise we first relate the above determinants to permutations and their signs.

Consider first the 2×2 case. The elements of S_2 are $\alpha = (1)$, and $\beta = (12)$, and sign $\alpha = +1$, sign $\beta = -1$. Moreover,

$$\det A = a_{11}a_{22} - a_{12}a_{21} = (\text{sign } \alpha)a_{1\alpha(1)}a_{2\alpha(2)} + (\text{sign } \beta)a_{1\beta(1)}a_{2\beta(2)}$$
$$= \sum_{\sigma \in S_2} (\text{sign } \sigma)a_{1\sigma(1)}a_{2\sigma(2)}.$$

What about the 3×3 case? The elements of S_3 are $\alpha = (1)$, $\beta = (123)$, $\gamma = (132)$, $\delta = (23)$, $\varepsilon = (12)$, $\mu = (13)$, and sign $\alpha = $ sign $\beta = $ sign $\gamma = +1$, sign $\delta = $ sign $\varepsilon = $ sign $\mu = -1$. Furthermore,

$$\det A = (\text{sign } \alpha)a_{1\alpha(1)}a_{2\alpha(2)}a_{3\alpha(3)} + (\text{sign } \beta)a_{1\beta(1)}a_{2\beta(2)}a_{3\beta(3)}$$
$$+ (\text{sign } \gamma)a_{1\gamma(1)}a_{2\gamma(2)}a_{3\gamma(3)} + (\text{sign } \delta)a_{1\delta(1)}a_{2\delta(2)}a_{3\delta(3)}$$
$$+ (\text{sign } \varepsilon)a_{1\varepsilon(1)}a_{2\varepsilon(2)}a_{3\varepsilon(3)} + (\text{sign } \mu)a_{1\mu(1)}a_{2\mu(2)}a_{3\mu(3)}$$
$$= \sum_{\sigma \in S_3} (\text{sign } \sigma)a_{1\sigma(1)}a_{2\sigma(2)}a_{3\sigma(3)}.$$

If in S_1 we put $\sigma = (1)$, then again

$$\det A = a_{11} = (\text{sign } \sigma)a_{1\sigma(1)} = \sum_{\sigma \in S_1} (\text{sign } \sigma)a_{1\sigma(1)}.$$

This suggests that we make the following definition. If $A = [a_{ij}]_{n \times n}$ then the *determinant* of A, $\det A$ (also written sometimes as $|A|$), is given by

$$\det A = \sum_{\sigma \in S_n} (\text{sign } \sigma)a_{1\sigma(1)}a_{2\sigma(2)} \ldots a_{n\sigma(n)}.$$

The determinant of A is often written as

$$\begin{vmatrix} a_{11} & a_{12} & \ldots & a_{1n} \\ a_{21} & a_{22} & \ldots & a_{2n} \\ \ldots & \ldots & \ldots & \ldots \\ a_{n1} & a_{n2} & \ldots & a_{nn} \end{vmatrix}.$$

We saw that if $A = [a_{ij}]_{n \times n}$ and $n = 1, 2$ or 3 then A is invertible precisely when $\det A \neq 0$. As our motivation for defining this number was to enable us to determine when a square matrix is invertible, we shall, naturally, judge the appropriateness of this generalisation by whether this result now holds for all positive integers n. As the definition of a determinant is somewhat involved it should come as no surprise that a little preliminary work is necessary before all the evidence is assembled!

Examples 4.2

1. $\begin{vmatrix} 1 & 2 \\ 3 & 4 \end{vmatrix} = 1 \times 4 - 2 \times 3 = 4 - 6 = -2.$

2. $\begin{vmatrix} 1 & -1 \\ 2 & -2 \end{vmatrix} = 1 \times (-2) - (-1) \times 2 = -2 + 2 = 0.$

For the moment we will not bother to practise the evaluation of larger determinants, as this task will be greatly eased by our later results. At the moment it appears rather daunting.

3. The matrix $A = [a_{ij}]_{n \times n}$ is called *upper triangular* if $a_{ij} = 0$ for $i > j$ (so that the only non-zero elements are on or above the main diagonal). Show that det $A = a_{11}a_{22} \ldots a_{nn}$ (the product of the elements on the main diagonal).

Solution If $\sigma \neq (1)$, then there is an i lying between 1 and n such that $\sigma(i) < i$ (see exercise 4.1.4). Since $a_{ij} = 0$ if $i > j$, $a_{i\sigma(i)} = 0$, and so $a_{1\sigma(1)} a_{2\sigma(2)} \ldots a_{n\sigma(n)} = 0$ for every σ apart from $\sigma = (1) = \sigma_1$, say.

Thus
$$\det A = (\text{sign } \sigma_1) a_{1\sigma_1(1)} a_{2\sigma_1(2)} \ldots a_{n\sigma_1(n)}$$
$$= a_{11}a_{22} \ldots a_{nn}.$$

EXERCISES 4.2

1 Evaluate the following determinants:

(a) $\begin{vmatrix} 1 & -1 \\ 0 & -2 \end{vmatrix}$; (b) $\begin{vmatrix} 3 & 5 \\ -1 & 2 \end{vmatrix}$; (c) $\begin{vmatrix} 4 & 1 \\ -1 & -2 \end{vmatrix}$;

(d) $\begin{vmatrix} \cos \theta & \sin \theta \\ -\sin \theta & \cos \theta \end{vmatrix}$.

2 The matrix $A = [a_{ij}]_{n \times n}$ is *lower triangular* if $a_{ij} = 0$ for $i < j$. Modify the argument used in example 4.2.3 in order to calculate det A.

3 Prove that an upper (or lower) triangular matrix is invertible if and only if det $A \neq 0$.

4.3 ELEMENTARY PROPERTIES OF DETERMINANTS

If we are to evaluate determinants successfully it will be helpful to accumulate some properties which they possess. Throughout the section, A will represent the $n \times n$ matrix $[a_{ij}]$.

THEOREM 4.3.1 $\det A = \sum_{\tau \in S_n} (\text{sign } \tau) a_{\tau(1)1} a_{\tau(2)2} \cdots a_{\tau(n)n}.$

Proof Suppose that $\sigma(r) = s$; then $r = \sigma^{-1}(s)$ and $a_{r\sigma(r)} = a_{\sigma^{-1}(s)s}.$

Hence $\det A = \sum_{\sigma \in S_n} (\text{sign } \sigma) a_{1\sigma(1)} a_{2\sigma(2)} \cdots a_{n\sigma(n)}$

$= \sum_{\sigma \in S_n} (\text{sign } \sigma) a_{\sigma^{-1}(1)1} a_{\sigma^{-1}(2)2} \cdots a_{\sigma^{-1}(n)n}$ by rearranging the $a_{i\sigma(i)}$s in each summand.

$= \sum_{\sigma \in S_n} (\text{sign } \sigma^{-1}) a_{\sigma^{-1}(1)1} a_{\sigma^{-1}(2)2} \cdots a_{\sigma^{-1}(n)n}.$

Now put $\tau = \sigma^{-1}$. Then, as σ runs through all elements of S_n, so does τ. Thus,

$$\det A = \sum_{\tau \in S_n} (\text{sign } \tau) a_{\tau(1)1} a_{\tau(2)2} \cdots a_{\tau(n)n}.$$

COROLLARY 4.3.2 $\det A = \det A^T$ (so that interchanging the rows and columns of a determinant does not alter its value).

Proof We have defined A^T to be $[b_{ij}]_{n \times n}$ where $b_{ij} = a_{ji}$. It follows that

$\det A^T = \sum_{\sigma \in S_n} (\text{sign } \sigma) b_{1\sigma(1)} b_{2\sigma(2)} \cdots b_{n\sigma(n)}$

$= \sum_{\sigma \in S_n} (\text{sign } \sigma) a_{\sigma(1)1} a_{\sigma(2)2} \cdots a_{\sigma(n)n}$

$= \det A$ by Theorem 4.3.1.

THEOREM 4.3.3 If every element of a row (or column) of A is multiplied by a real number λ then $\det A$ is multiplied by λ.

Proof Suppose that row r of A is multiplied by λ to give matrix $B = [b_{ij}]_{n \times n}$ so that

$$b_{ij} = \begin{cases} a_{ij} & (i \neq r) \\ \lambda a_{ij} & (i = r). \end{cases}$$

Then $\det B = \sum_{\sigma \in S_n} (\text{sign } \sigma) b_{1\sigma(1)} \cdots b_{r\sigma(r)} \cdots b_{n\sigma(n)}$

$$= \sum_{\sigma \in S_n} (\text{sign } \sigma) a_{1\sigma(1)} \ldots (\lambda a_{r\sigma(r)}) \ldots a_{n\sigma(n)}$$

$$= \lambda \sum_{\sigma \in S_n} (\text{sign } \sigma) a_{1\sigma(1)} \ldots a_{r\sigma(r)} \ldots a_{n\sigma(n)} = \lambda \det A.$$

Multiplication of columns can be dealt with similarly.

COROLLARY 4.3.4 $\det(\lambda A) = \lambda^n \det A$.

COROLLARY 4.3.5 If every element of a row (or column) of A is zero, then $\det A = 0$.

Proof Suppose that every element of row r is zero. Then multiplying row r by 0 we see that $\det A = 0 \det A = 0$.

THEOREM 4.3.6 Interchanging two rows (or columns) of A changes the sign of $\det A$.

Proof Suppose that rows r and s are interchanged in A to produce the matrix $B = [b_{ij}]$ where

$$b_{ij} = \begin{cases} a_{ij} & (i \neq r, s) \\ a_{rj} & (i = s) \\ a_{sj} & (i = r), \end{cases}$$

and let $\psi = (rs)$. Then

$$\det B = \sum_{\sigma \in S_n} (\text{sign } \sigma) b_{1\sigma(1)} \ldots b_{r\sigma(r)} \ldots b_{s\sigma(s)} \ldots b_{n\sigma(n)}$$

$$= \sum_{\sigma \in S_n} (\text{sign } \sigma) a_{1\sigma(1)} \ldots a_{s\sigma(r)} \ldots a_{r\sigma(s)} \ldots a_{n\sigma(n)}$$

$$= \sum_{\sigma \in S_n} (\text{sign } \sigma) a_{1\sigma\psi(1)} \ldots a_{s\sigma\psi(s)} \ldots a_{r\sigma\psi(r)} \ldots a_{n\sigma\psi(n)}$$

since $\sigma\psi(i) = \begin{cases} \sigma(i) & (i \neq r, s) \\ \sigma(s) & (i = r) \\ \sigma(r) & (i = s) \end{cases}$

$$= \sum_{\sigma\psi \in S_n} (\text{sign } \sigma) a_{1\sigma\psi(1)} \ldots a_{s\sigma\psi(s)} \ldots a_{r\sigma\psi(r)} \ldots a_{n\sigma\psi(n)},$$

because, as σ runs through the whole of S_n, so does $\sigma\psi$.

$$\therefore \quad \det B = \sum_{\tau \in S_n} (\text{sign } \sigma) a_{1\tau(1)} a_{2\tau(2)} \ldots a_{n\tau(n)}, \quad \text{putting } \tau = \sigma\psi;$$

$$= -\sum_{\tau \in S_n} (\text{sign } \tau) a_{1\tau(1)} a_{2\tau(2)} \ldots a_{n\tau(n)}, \quad \text{since sign } \tau = \text{sign}(\sigma\psi)$$
$$= \text{sign } \sigma \text{ sign } \psi = -\text{sign } \sigma.$$

$$= -\det A.$$

The interchange of two columns can be treated similarly.

COROLLARY 4.3.7 If two rows (or columns) of A are identical, then $\det A = 0$.

Proof Let B be the matrix obtained from A by interchanging the two identical rows. Then $\det B = -\det A$ by Theorem 4.3.6. But, clearly, $B = A$, and so $\det A = 0$.

THEOREM 4.3.8 Let

$$A = \begin{bmatrix} a_{11} & \cdots & a_{1n} \\ \cdots & & \cdots \\ b_{i1} + c_{i1} & \cdots & b_{in} + c_{in} \\ \cdots & & \cdots \\ a_{n1} & \cdots & a_{nn} \end{bmatrix}, \quad B = \begin{bmatrix} a_{11} & \cdots & a_{1n} \\ \cdots & & \cdots \\ b_{i1} & \cdots & b_{in} \\ \cdots & & \cdots \\ a_{n1} & \cdots & a_{nn} \end{bmatrix}, \quad C = \begin{bmatrix} a_{11} & \cdots & a_{1n} \\ \cdots & & \cdots \\ c_{i1} & \cdots & c_{in} \\ \cdots & & \cdots \\ a_{n1} & \cdots & a_{nn} \end{bmatrix}.$$

Then $\det A = \det B + \det C$. This holds for any i such that $1 \leqslant i \leqslant n$; a similar result holds for columns.

Proof
$$\det A = \sum_{\sigma \in S_n} (\text{sign } \sigma) a_{1\sigma(1)} \ldots (b_{i\sigma(i)} + c_{i\sigma(i)}) \ldots a_{n\sigma(n)}$$

$$= \sum_{\sigma \in S_n} (\text{sign } \sigma) a_{1\sigma(1)} \ldots b_{i\sigma(i)} \ldots a_{n\sigma(n)}$$

$$+ \sum_{\sigma \in S_n} (\text{sign } \sigma) a_{1\sigma(1)} \ldots c_{i\sigma(i)} \ldots a_{n\sigma(n)}$$

$$= \det B + \det C.$$

COROLLARY 4.3.9 Let $\lambda \in \mathbb{R}$. Then, adding $\lambda \times$ row s to row r ($r \neq s$) of A leaves $\det A$ unchanged. Again, a similar result holds for columns.

Proof

$$\text{Row } r \to \begin{vmatrix} a_{11} & \cdots & a_{1n} \\ \cdots & \cdots & \cdots \\ a_{r1}+\lambda a_{s1} & \cdots & a_{rn}+\lambda a_{sn} \\ \cdots & \cdots & \cdots \\ a_{n1} & \cdots & a_{nn} \end{vmatrix}$$

$$= \begin{vmatrix} a_{11} & \cdots & a_{1n} \\ \cdots & \cdots & \cdots \\ a_{r1} & \cdots & a_{rn} \\ \cdots & \cdots & \cdots \\ a_{n1} & \cdots & a_{nn} \end{vmatrix} + \begin{vmatrix} a_{11} & \cdots & a_{1n} \\ \cdots & \cdots & \cdots \\ \lambda a_{s1} & \cdots & \lambda a_{sn} \\ \cdots & \cdots & \cdots \\ a_{n1} & \cdots & a_{nn} \end{vmatrix} \quad \text{by Theorem 4.3.8}$$

$$= \det A + \lambda \begin{vmatrix} a_{11} & \cdots & a_{1n} \\ \cdots & \cdots & \cdots \\ a_{s1} & \cdots & a_{sn} & \leftarrow \text{row } r \\ \cdots & \cdots & \cdots \\ a_{s1} & \cdots & a_{sn} & \leftarrow \text{row } s \\ \cdots & \cdots & \cdots \\ a_{n1} & \cdots & a_{nn} \end{vmatrix} \quad \text{by Theorem 4.3.3}$$

$$= \det A + \lambda 0 \quad \text{by Corollary 4.3.7}$$
$$= \det A.$$

Summary

(a) Interchanging the rows and columns of a determinant does not alter its value.
(b) If every element of a row of A is multiplied by a real number λ then $\det A$ is multiplied by λ.
(c) If every element of a row of A is zero then $\det A = 0$.
(d) Interchanging two rows of A changes the sign of $\det A$.
(e) If two rows of A are identical then $\det A = 0$.
(f) Adding a multiple of one row of A to another row of A leaves $\det A$ unchanged.

In each of the above we can replace 'row(s)' by 'column(s)'.
These results allow us to simplify the determinant of a square matrix A, or to evaluate it by reducing A to echelon form.

Examples 4.3

1. $\begin{vmatrix} 1 & 2 & 3 \\ 4 & 5 & 6 \\ 7 & 8 & 9 \end{vmatrix} = \begin{vmatrix} 1 & 2 & 3 \\ 0 & -3 & -6 \\ 0 & -6 & -12 \end{vmatrix}$ $(R_2 = r_2 - 4r_1)$ by Corollary 4.3.9
    $(R_3 = r_3 - 7r_1)$

    $= 2 \begin{vmatrix} 1 & 2 & 3 \\ 0 & -3 & -6 \\ 0 & -3 & -6 \end{vmatrix}$ $(R_3 = \tfrac{1}{2}r_3)$ by Theorem 4.3.3

    $= 0$ by Corollary 4.3.7

2. $\begin{vmatrix} 1 & 1 & 1 \\ 1 & -1 & 1 \\ 2 & 3 & -1 \end{vmatrix} = \begin{vmatrix} 1 & 1 & 1 \\ 0 & -2 & 0 \\ 0 & 1 & -3 \end{vmatrix}$ $(R_2 = r_2 - r_1)$ by Corollary 4.3.9
    $(R_3 = r_3 - 2r_1)$

    $= 2 \begin{vmatrix} 1 & 1 & 1 \\ 0 & -1 & 0 \\ 0 & 1 & -3 \end{vmatrix}$ $(R_2 = \tfrac{1}{2}r_2)$ by Theorem 4.3.3

    $= 2 \begin{vmatrix} 1 & 1 & 1 \\ 0 & -1 & 0 \\ 0 & 0 & -3 \end{vmatrix}$ by Corollary 4.3.9
    $(R_3 = r_3 + r_2)$

    $= 2 \times 1 \times (-1) \times (-3)$ by Example 4.2.3

    $= 6.$

EXERCISES 4.3

1 Evaluate the following determinants:

(a) $\begin{vmatrix} 8 & 3 & 4 \\ 6 & 4 & 5 \\ 10 & 5 & 6 \end{vmatrix}$;   (b) $\begin{vmatrix} 1 & 2 & 3 \\ 2 & 3 & 4 \\ 3 & 4 & 5 \end{vmatrix}$;   (c) $\begin{vmatrix} 7 & 10 & 5 \\ 13 & 6 & 2 \\ 11 & 14 & 8 \end{vmatrix}$.

2 Find the determinants of the elementary matrices $E_r(\alpha)$, $E_{rs}(\alpha)$ and E_{rs}. In particular, verify that they are all non-zero.

3 Let A be any $n \times n$ matrix.

(a) If E is an elementary matrix of size n, prove that

$$\det(EA) = (\det E)(\det A) = \det(AE).$$

(b) If $E_1, \ldots, E_m$ are elementary matrices of size n, prove that

$$\det(E_1 \ldots E_m A) = (\det E_1) \ldots (\det E_m)(\det A) = \det(A E_1 \ldots E_m).$$

4 Find two 2×2 real matrices A, B such that $\det(A + B) \neq \det A + \det B$.

4.4 NON-SINGULAR MATRICES

Throughout this section A denotes the $n \times n$ matrix $[a_{ij}]$. It was shown in Theorem 3.4.1 that A is row-equivalent to a matrix B in reduced echelon form. Moreover, we saw in section 3.5 that we can write A as

$$A = E_1 \ldots E_k B \quad \text{where } E_1, \ldots, E_k \text{ are elementary matrices.}$$

There are only two possible forms for B, as was pointed out in exercise 3.6.2, namely,

(a) $B = I_n$, the $n \times n$ identity matrix if and only if A is invertible, and
(b) B has at least one row of zeros if and only if A is not invertible.

Now $\det A = \det(E_1 \ldots E_k B)$

$= (\det E_1) \ldots (\det E_k)(\det B) \quad$ by exercise 4.3.3.

If $B = I_n$, then $\det A = (\det E_1) \ldots (\det E_k) \neq 0$ (exercise 4.3.2); if B has a row of zeros, then $\det B = 0$ (Corollary 4.3.5), so that $\det A = 0$. It follows that A is invertible if and only if $\det A \neq 0$. A matrix A for which $\det A = 0$ is usually called *singular*; correspondingly, if $\det A \neq 0$ then A is termed *non-singular*. We have thus proved the following:

THEOREM 4.4.1 The $n \times n$ matrix A is invertible if and only if it is non-singular.

This provides us with an answer to question (b) on p. 44. Of course, we have side-stepped the issue of the practicality of calculating the determinant of a square matrix, but the neatness of the criterion is appealing.

We can also use the ideas expounded above to prove the following useful result.

THEOREM 4.4.2 Let $A = [a_{ij}]_{n \times n}$, $B = [b_{ij}]_{n \times n}$ be two square matrices. Then $\det(AB) = (\det A)(\det B)$.

Proof Let $A = E_1 \ldots E_k C$ where $E_1, \ldots, E_k$ are elementary matrices and C is the reduced echelon form of A. There are two possibilities to consider:

(a) $C = I_n$, in which case $A = E_1 \ldots E_k$ and, using exercise 4.3.3(b),

$$\det(AB) = \det(E_1 \ldots E_k B) = (\det E_1) \ldots (\det E_k)(\det B)$$
$$= \det(E_1 \ldots E_k)(\det B) = (\det A)(\det B).$$

(b) C has a row of zeros, in which case CB has a row of zeros (see exercise 2.4.7), so $\det C = 0$ and $\det(CB) = 0$.

Then $\det A = \det(E_1 \ldots E_k C) = (\det E_1) \ldots (\det E_k)(\det C) = 0$.

Similarly,
$$\det(AB) = \det(E_1 \ldots E_k CB) = (\det E_1) \ldots (\det E_k)[\det(CB)] = 0.$$

Hence, $(\det A)(\det B) = 0 = \det(AB)$.

COROLLARY 4.4.3 If A is invertible then $\det(A^{-1}) = 1/\det A$.

Proof Let A be invertible, so that $AA^{-1} = I_n$.

Then $(\det A)(\det(A^{-1})) = \det(AA^{-1}) = \det I_n = 1$,

so that $\det(A^{-1}) = 1/\det A$.

Example 4.4

For what values of x is the matrix $\begin{vmatrix} 1 & 2 & -1 \\ x & 1 & 3 \\ 3 & 3 & x \end{vmatrix}$ invertible?

Solution

$$\begin{vmatrix} 1 & 2 & -1 \\ x & 1 & 3 \\ 3 & 3 & x \end{vmatrix} = \begin{vmatrix} 1 & 2 & -1 \\ 0 & 1-2x & 3+x \\ 0 & -3 & x+3 \end{vmatrix} \quad \begin{matrix} (R_2 = r_2 - xr_1) \\ (R_3 = r_3 - 3r_1) \end{matrix} \quad \text{by Corollary 4.3.9}$$

$$= \begin{vmatrix} 1 & 2 & -1 \\ 0 & 1-2x & 3+x \\ 0 & 2x-4 & 0 \end{vmatrix} \quad (R_3 = r_3 - r_2) \quad \text{by Corollary 4.3.9}$$

$$= - \begin{vmatrix} 1 & -1 & 2 \\ 0 & 3+x & 1-2x \\ 0 & 0 & 2x-4 \end{vmatrix} \quad \begin{matrix} (C_2 = c_3) \\ (C_3 = c_2) \end{matrix} \quad \text{by Theorem 4.3.6}$$

$$= -(3+x)(2x-4) \quad \text{by example 4.2.3.}$$

Hence the matrix is invertible if and only if $x \neq 2, -3$.

EXERCISES 4.4

1. For what real values of a is the matrix $\begin{bmatrix} 1-a & 1 & -1 \\ 2 & 1-a & 2 \\ 2 & -1 & 4-a \end{bmatrix}$ invertible?

2. Let A, B be square matrices. Prove that if AB is invertible then both A and B are invertible.

3. Let A, B be $n \times n$ matrices. Prove that if B is invertible then
$$\det(I_n - B^{-1}AB) = \det(I_n - A).$$

4. Let A be a $(2n+1) \times (2n+1)$ matrix. Prove that if $A^T = -A$ then A is singular.

4.5 COFACTORS

The results of the previous sections certainly help us to manipulate determinants, but we still have no systematic method for evaluating them. The main purpose here is to develop an inductive procedure for finding the determinant of any given square matrix.

Let $A = [a_{ij}]_{n \times n}$ where $n > 1$. For r, s such that $1 \leqslant r, s \leqslant n$ we let Δ_{rs} be the determinant of the $(n-1) \times (n-1)$ matrix obtained from A by omitting row r and column s, and put $A_{rs} = (-1)^{r+s}\Delta_{rs}$. Then Δ_{rs} is known as the (r, s)-*minor* of A; A_{rs} is the (r, s)-*cofactor* of A.

Example 4.5.1

If $A = \begin{vmatrix} 1 & 2 & 3 \\ 4 & 5 & 6 \\ 7 & 8 & 9 \end{vmatrix}$ then $A_{11} = (-1)^{1+1}\Delta_{11} = \begin{vmatrix} 5 & 6 \\ 8 & 9 \end{vmatrix} = 45 - 48 = -3,$

$$A_{12} = (-1)^{1+2}\Delta_{12} = (-1)\begin{vmatrix} 4 & 6 \\ 7 & 9 \end{vmatrix} = -(36 - 42) = 6,$$

$$A_{23} = (-1)^{2+3}\Delta_{23} = (-1)\begin{vmatrix} 1 & 2 \\ 7 & 8 \end{vmatrix} = -(8 - 14) = 6.$$

LEMMA 4.5.1

Let $A = \begin{vmatrix} 1 & 0 & \ldots & 0 \\ a_{21} & a_{22} & \ldots & a_{2n} \\ \vdots & & & \\ a_{n1} & a_{n2} & \ldots & a_{nn} \end{vmatrix}$, $B = \begin{vmatrix} a_{22} & \ldots & a_{2n} \\ \vdots & & \\ a_{n2} & \ldots & a_{nn} \end{vmatrix}$.

Then $\det A = \det B$.

Proof Let T_n be the set of all permutations of $2, \ldots, n$. For each $\psi \in T_n$ let $\sigma \in S_n$ be defined by

$$\sigma(i) = \begin{cases} 1 & \text{if } i = 1 \\ \psi(i) & \text{if } i > 1. \end{cases}$$

Then $\operatorname{sign} \sigma = \operatorname{sign} \psi$ (see exercise 4.1.5) and

$$\det A = \sum_{\sigma \in S_n} (\operatorname{sign} \sigma) a_{1\sigma(1)} a_{2\sigma(2)} \ldots a_{n\sigma(n)}$$

$$= \sum_{\substack{\sigma \in S_n \\ \text{such that } \sigma(1) = 1}} (\operatorname{sign} \sigma) a_{1\sigma(1)} a_{2\sigma(2)} \ldots a_{n\sigma(n)}, \quad \text{since } a_{1j} = 0 \text{ if } j > 1$$

$$= \sum_{\psi \in T_n} (\operatorname{sign} \psi) a_{2\psi(2)} \ldots a_{n\psi(n)} = \det B.$$

LEMMA 4.5.2 Let $n > 1$. Then

$$\begin{vmatrix} a_{11} & \ldots & a_{1(s-1)} & a_{1s} & a_{1(s+1)} & \ldots & a_{1n} \\ \vdots & & & & & & \vdots \\ 0 & \ldots & 0 & 1 & 0 & \ldots & 0 \\ \vdots & & & & & & \vdots \\ a_{n1} & \ldots & a_{n(s-1)} & a_{ns} & a_{n(s+1)} & \ldots & a_{nn} \end{vmatrix} = A_{rs}$$

row $r \to$ (indicates the middle row), column s (indicates the column with the 1).

Proof

$$\begin{vmatrix} a_{11} & \ldots & a_{1(s-1)} & a_{1s} & a_{1(s+1)} & \ldots & a_{1n} \\ \vdots & & & & & & \vdots \\ 0 & \ldots & 0 & 1 & 0 & \ldots & 0 \\ \vdots & & & & & & \vdots \\ a_{n1} & \ldots & a_{n(s-1)} & a_{ns} & a_{n(s+1)} & \ldots & a_{nn} \end{vmatrix}$$

row $r \to$, column s

$$= (-1)^{r-1} \begin{vmatrix} 0 & \ldots & 0 & 1 & 0 & \ldots & 0 \\ a_{11} & \ldots & a_{1(s-1)} & a_{1s} & a_{1(s+1)} & \ldots & a_{1n} \\ \vdots & & & & & & \vdots \\ a_{n1} & \ldots & a_{n(s-1)} & a_{ns} & a_{n(s+1)} & \ldots & a_{nn} \end{vmatrix}$$

(by using Theorem 4.3.6 $(r-1)$ times; that is, by interchanging rows r and $r-1$, then $r-1$ and $r-2$, and so on).

$$= (-1)^{r-1}(-1)^{s-1} \begin{vmatrix} 1 & 0 & \ldots & 0 & 0 & \ldots & 0 \\ a_{1s} & a_{11} & \ldots & a_{1(s-1)} & a_{1(s+1)} & \ldots & a_{1n} \\ \vdots & & & & & & \vdots \\ a_{ns} & a_{n1} & \ldots & a_{n(s-1)} & a_{n(s+1)} & \ldots & a_{nn} \end{vmatrix}$$

(using Theorem 4.3.6 again, this time applied to column interchanges)

$$= (-1)^{r+s-2} \Delta_{rs} \quad \text{by Lemma 4.5.1}$$
$$= (-1)^{r+s} \Delta_{rs} = A_{rs}.$$

THEOREM 4.5.3 Let $A = [a_{ij}]_{n \times n}$ $(n > 1)$ and let r be such that $1 \leq r \leq n$. Then
$$\det A = a_{r1}A_{r1} + a_{r2}A_{r2} + \ldots + a_{rn}A_{rn} \left(= \sum_{s=1}^{n} a_{rs}A_{rs} \right) \quad (1)$$

and
$$\det A = a_{1r}A_{1r} + a_{2r}A_{2r} + \ldots + a_{nr}A_{nr} \left(= \sum_{s=1}^{n} a_{sr}A_{sr} \right). \quad (2)$$

Proof

$$\begin{vmatrix} a_{11} & \ldots & a_{1n} \\ \vdots & & \vdots \\ a_{r1} & \ldots & a_{rn} \\ \vdots & & \vdots \\ a_{n1} & \ldots & a_{nn} \end{vmatrix} = \begin{vmatrix} a_{11} & a_{12} & \ldots & a_{1n} \\ \vdots & & & \vdots \\ a_{r1} & 0 & \ldots & 0 \\ \vdots & & & \vdots \\ a_{n1} & a_{n2} & \ldots & a_{nn} \end{vmatrix} + \begin{vmatrix} a_{11} & a_{12} & \ldots & a_{1n} \\ \vdots & & & \vdots \\ 0 & a_{r2} & \ldots & 0 \\ \vdots & & & \vdots \\ a_{n1} & a_{n2} & \ldots & a_{nn} \end{vmatrix}$$

$$+ \ldots + \begin{vmatrix} a_{11} & \ldots & a_{1(n-1)} & a_{1n} \\ \vdots & & & \vdots \\ 0 & \ldots & 0 & a_{rn} \\ \vdots & & & \vdots \\ a_{n1} & \ldots & a_{n(n-1)} & a_{nn} \end{vmatrix}$$

(by Theorem 4.3.8)

$$= a_{r1} \begin{vmatrix} a_{11} & a_{12} & \cdots & a_{1n} \\ \cdots\cdots\cdots\cdots\cdots\cdots \\ 1 & 0 & \cdots & 0 \\ \cdots\cdots\cdots\cdots\cdots\cdots \\ a_{n1} & a_{n2} & \cdots & a_{nn} \end{vmatrix} + a_{r2} \begin{vmatrix} a_{11} & a_{12} & \cdots & a_{1n} \\ \cdots\cdots\cdots\cdots\cdots\cdots \\ 0 & 1 & \cdots & 0 \\ \cdots\cdots\cdots\cdots\cdots\cdots \\ a_{n1} & a_{n2} & \cdots & a_{nn} \end{vmatrix}$$

$$+ \ldots + a_{rn} \begin{vmatrix} a_{11} & \cdots & a_{1(n-1)} & a_{1n} \\ \cdots\cdots\cdots\cdots\cdots\cdots\cdots \\ 0 & \cdots & 0 & 1 \\ \cdots\cdots\cdots\cdots\cdots\cdots\cdots \\ a_{n1} & \cdots & a_{n(n-1)} & a_{nn} \end{vmatrix}$$

$$= a_{r1}A_{r1} + a_{r2}A_{r2} + \ldots + a_{rn}A_{rn} \text{ by Lemma 4.5.2.}$$

Equation (2) can be verified in a similar way.

Equation (1) (respectively (2)) in Theorem 4.5.3 above is called the *expansion of* det A *by the rth row* (respectively *rth column*) *of* A.

Given a particular determinant to evaluate, we would normally use a combination of the techniques in sections 4.3 and 4.5.

Examples 4.5.2

1. $\begin{vmatrix} 1 & 2 & -2 \\ -1 & 3 & 0 \\ 0 & -2 & 1 \end{vmatrix} = \begin{vmatrix} 1 & 2 & -2 \\ 0 & 5 & -2 \\ 0 & -2 & 1 \end{vmatrix}$ $(R_2 = r_2 + r_1)$ by Corollary 4.3.9

    $\qquad\qquad\qquad = \begin{vmatrix} 5 & -2 \\ -2 & 1 \end{vmatrix}$ expanding by the first column

    $\qquad\qquad\qquad = 5 - 4 = 1.$

2. Express $\begin{vmatrix} 1 & 1 & 1 \\ a^2 & b^2 & c^2 \\ a^3 & b^3 & c^3 \end{vmatrix}$ as a product of linear and quadratic factors.

    Solution   The required determinant $= \begin{vmatrix} 1 & 0 & 0 \\ a^2 & b^2 - a^2 & c^2 - a^2 \\ a^3 & b^3 - a^3 & c^3 - a^3 \end{vmatrix}$ $\begin{matrix}(C_2 = c_2 - c_1)\\(C_3 = c_3 - c_1)\end{matrix}$

    $= \begin{vmatrix} b^2 - a^2 & c^2 - a^2 \\ b^3 - a^3 & c^3 - a^3 \end{vmatrix}$ expanding by the first row

$$= (b-a)(c-a) \begin{vmatrix} b+a & c+a \\ b^2+ba+a^2 & c^2+ca+a^2 \end{vmatrix}$$

$$= (b-a)(c-a) \begin{vmatrix} b+a & c+a \\ b^2 & c^2 \end{vmatrix} \quad (R_2 = r_2 - ar_1)$$

$$= (b-a)(c-a) \begin{vmatrix} b+a & c-b \\ b^2 & c^2-b^2 \end{vmatrix} \quad (c_2 = c_2 - c_1)$$

$$= (b-a)(c-a)(c-b) \begin{vmatrix} b+a & 1 \\ b^2 & c+b \end{vmatrix}$$

$$= (b-a)(c-a)(c-b)(bc+ca+ab).$$

EXERCISES 4.5

1 Evaluate the following determinants:

(a) $\begin{vmatrix} 1 & 1 & -2 \\ 3 & -1 & -6 \\ -2 & 3 & 4 \end{vmatrix}$; (b) $\begin{vmatrix} 4 & 2 & 2 \\ 4 & 2 & -1 \\ -1 & 3 & 7 \end{vmatrix}$;

(c) $\begin{vmatrix} 5 & 6 & 8 & -1 \\ 4 & 3 & 0 & 0 \\ 10 & 12 & 16 & -2 \\ 1 & 2 & 0 & 0 \end{vmatrix}$.

2 Prove that $\begin{vmatrix} a-x & a-y & a-z \\ b-x & b-y & b-z \\ c-x & c-y & c-z \end{vmatrix} = 0.$

3 Let D_n denote $n \times n$ determinant

$$\begin{vmatrix} 1+x^2 & x & 0 & \cdots & 0 & 0 & 0 \\ x & 1+x^2 & x & \cdots & 0 & 0 & 0 \\ 0 & x & 1+x^2 & \cdots & 0 & 0 & 0 \\ \multicolumn{7}{c}{\cdots\cdots\cdots\cdots\cdots\cdots\cdots\cdots\cdots\cdots\cdots\cdots\cdots\cdots} \\ 0 & 0 & 0 & \cdots & 0 & x & 1+x^2 \end{vmatrix}.$$

Prove that $D_n - D_{n-1} = x^2(D_{n-1} - D_{n-2})$, and hence evaluate D_n.

4 Prove that $\begin{vmatrix} a & -b & -a & b \\ b & a & -b & -a \\ c & -d & c & -d \\ d & c & d & c \end{vmatrix} = 4(a^2+b^2)(c^2+d^2).$

5 By considering the product of $\begin{vmatrix} a^2 & a & 1 \\ b^2 & b & 1 \\ c^2 & c & 1 \end{vmatrix}$ and $\begin{vmatrix} 1 & -2x & x^2 \\ 1 & -2y & y^2 \\ 1 & -2z & z^2 \end{vmatrix}$

express $\begin{vmatrix} (a-x)^2 & (a-y)^2 & (a-z)^2 \\ (b-x)^2 & (b-y)^2 & (b-z)^2 \\ (c-x)^2 & (c-y)^2 & (c-z)^2 \end{vmatrix}$ as a product of linear factors.

6 By considering the product of two determinants of the form
$\begin{vmatrix} a+ib & c+id \\ -(c-id) & a-ib \end{vmatrix}$ over the complex numbers, prove that the product of a sum of four squares by a sum of four squares is itself a sum of four squares.

4.6 THE ADJUGATE OF A MATRIX

The theory of determinants which we have presented can be used to give an alternative method for finding the inverse of an invertible matrix. Despite the fact that the method is of little practical value, it does have some theoretical interest, and so we will spend a little time revealing it.

THEOREM 4.6.1 Let $A = [a_{ij}]_{n \times n}$ ($n > 1$), and let A_{ij} be the (i,j)-cofactor of A. Then, if $r \neq s$,

(a) $a_{r1}A_{s1} + a_{r2}A_{s2} + \ldots + a_{rn}A_{sn}\left(= \sum_{t=1}^{n} a_{rt}A_{st} \right) = 0$, and

(b) $a_{1r}A_{1s} + a_{2r}A_{2s} + \ldots + a_{nr}A_{ns}\left(= \sum_{t=1}^{n} a_{tr}A_{ts} \right) = 0.$

Proof

(a) Suppose that $r \neq s$ and put $B = \begin{bmatrix} a_{11} \ldots a_{1n} \\ \ldots\ldots\ldots \\ a_{r1} \ldots a_{rn} \\ \ldots\ldots\ldots \\ a_{r1} \ldots a_{rn} \\ \ldots\ldots\ldots \\ a_{n1} \ldots a_{nn} \end{bmatrix}$ ← row r

← row s

Expanding by row s, we get $\det B = a_{r1}A_{s1} + \ldots + a_{rn}A_{sn}$. But $\det B = 0$ by Corollary 4.3.7, so the result follows.

(b) This can be shown similarly.

THEOREM 4.6.2 Let $A = [a_{ij}]_{n \times n}$ ($n > 1$), and let $B = [A_{ij}]_{n \times n}$ where A_{ij} is the (i,j)-cofactor of A. Then
$$AB^T = B^T A = (\det A) I_n.$$

Proof Let $B^T = [b_{ij}]_{n \times n}$, so that $b_{ij} = A_{ji}$. Then the (r,s)-element of AB^T is

$a_{r1}b_{1s} + a_{r2}b_{2s} + \ldots + a_{rn}b_{ns}$ (by the definition of matrix product)

$= a_{r1}A_{s1} + a_{r2}A_{s2} + \ldots + a_{rn}A_{sn}$

$= \begin{cases} \det A & \text{if } r = s \quad \text{(Theorem 4.5.3)} \\ 0 & \text{if } r \neq s \quad \text{(Theorem 4.6.1).} \end{cases}$

Hence $AB^T = \begin{bmatrix} \det A & 0 & \ldots & 0 \\ 0 & \det A & \ldots & 0 \\ \multicolumn{4}{c}{\ldots\ldots\ldots\ldots\ldots\ldots} \\ 0 & 0 & \ldots & \det A \end{bmatrix} = (\det A) I_n$

The proof that $B^T A = (\det A) I_n$ is similar.

The matrix B^T of Theorem 4.6.2 is called the *adjugate* of A, and is denoted by adj A. We have seen that

$$A(\text{adj } A) = (\text{adj } A)A = (\det A) I_n.$$

This leads to the following corollary.

COROLLARY 4.6.3 If A is invertible, then $A^{-1} = \dfrac{1}{\det A}(\text{adj } A)$.

Proof If A is invertible, then $\det A \neq 0$, by Theorem 4.4.1, and so
$$A[(\det A)^{-1}(\text{adj } A)] = [(\det A)^{-1}(\text{adj } A)]A = I_n,$$
whence the result.

Examples 4.6

1. Let $A = \begin{bmatrix} 1 & 2 & 3 \\ 2 & 3 & 1 \\ 3 & 2 & 1 \end{bmatrix}$. Then adj $A = \begin{bmatrix} 1 & 4 & -7 \\ 1 & -8 & 5 \\ -5 & 4 & -1 \end{bmatrix}$

and $\det A = \begin{vmatrix} 1 & 2 & 3 \\ 0 & -1 & -5 \\ 0 & -4 & -8 \end{vmatrix} = \begin{vmatrix} -1 & -5 \\ -4 & -8 \end{vmatrix} = 8 - 20 = -12.$

Hence $A^{-1} = (-\frac{1}{12}) \begin{bmatrix} 1 & 4 & -7 \\ 1 & -8 & 5 \\ -5 & 4 & -1 \end{bmatrix}$.

2. Let A be an $n \times n$ non-singular matrix. Prove that $\operatorname{adj}(\alpha A) = \alpha^{n-1}(\operatorname{adj} A)$ for all real numbers α.

Solution By Theorem 4.6.2, $(\alpha A)\operatorname{adj}(\alpha A) = \det(\alpha A)I_n = \alpha^n(\det A)I_n$. Multiplying both sides by $\operatorname{adj} A$ gives $A(\operatorname{adj} A)(\operatorname{adj}(\alpha A)) = \alpha^{n-1}(\det A)(\operatorname{adj} A)$. Thus $(\det A)(\operatorname{adj}(\alpha A)) = \alpha^{n-1}(\det A)(\operatorname{adj} A)$. Since A is non-singular, we can cancel $\det A$ from both sides to give the required result.

EXERCISES 4.6

1 Compute the adjugate of the matrix $A = \begin{bmatrix} 1 & 2 & -2 \\ -1 & 3 & 0 \\ 0 & -2 & 1 \end{bmatrix}$,

and hence find A^{-1}.

2 Let A, B be $n \times n$ non-singular matrices. Prove the following:

(a) $\operatorname{adj}(A^{-1}) = (\operatorname{adj} A)^{-1}$;

(b) $\operatorname{adj}(AB) = (\operatorname{adj} B)(\operatorname{adj} A)$;

(c) $\operatorname{adj}(\operatorname{adj}(\operatorname{adj} A)) = (\det A)^{n^2 - 3n + 3} A^{-1}$.

3 (a) Prove that if A is a singular $n \times n$ matrix, then $\operatorname{adj} A$ is singular;

(b) Using (a), or otherwise, show that $\det(\operatorname{adj} A) = (\det A)^{n-1}$ for *all* $n \times n$ matrices A.

4 Let $\omega(\neq 1)$ be a complex cube root of 1. Prove that $1 + \omega + \omega^2 = 0$. Hence determine the inverse of $\begin{bmatrix} 1 & 1 & 1 \\ 1 & \omega & \omega^2 \\ 1 & \omega^2 & \omega \end{bmatrix}$.

4.7 SYSTEMS OF HOMOGENEOUS LINEAR EQUATIONS

Consider the following system of linear equations:
$$a_{11}x_1 + a_{12}x_2 + \ldots + a_{1n}x_n = 0,$$
$$a_{21}x_1 + a_{22}x_2 + \ldots + a_{2n}x_n = 0,$$
$$\ldots\ldots\ldots\ldots\ldots\ldots\ldots\ldots\ldots\ldots,$$
$$a_{m1}x_1 + a_{m2}x_2 + \ldots + a_{mn}x_n = 0.$$

Linear equations such as these, in which the right-hand side is zero, are referred to as *homogeneous*. As we saw in section 2.3, we can write this system in matrix notation as $AX = O$,

where $A = \begin{bmatrix} a_{11} \cdots a_{1n} \\ \cdots \cdots \\ a_{m1} \cdots a_{mn} \end{bmatrix}$, $X = \begin{bmatrix} x_1 \\ \vdots \\ x_n \end{bmatrix}$, $O = \begin{bmatrix} 0 \\ \vdots \\ 0 \end{bmatrix}$

and O is $m \times 1$. Notice that the system always has the *trivial* solution $x_1 = x_2 = \ldots = x_n = 0$.

THEOREM 4.7.1 Let $A = [a_{ij}]_{n \times n}$ ($m = n$, notice!). Then the system $AX = O$ has a non-trivial solution if and only if $\det A = 0$.

Proof Assume first that $\det A \neq 0$, so that A is invertible. Let $X = U$ be a solution to the system. Then $AU = O$ and so $U = A^{-1}AU = A^{-1}O = O$. Thus the system has only the trivial solution.

It remains to show that there is a non-trivial solution if $\det A = 0$. We prove this by induction on n. If $n = 1$ the system is $a_{11}x_1 = 0$, where $a_{11} = 0$ because $\det A = 0$. Clearly, any non-zero real value for x_1 is a non-trivial solution to this equation. So let $n > 1$, assume the result holds for systems whose matrix of coefficients is $(n-1) \times (n-1)$, and let A be $n \times n$ and singular.

If $a_{i1} = 0$ for all i such that $1 \leq i \leq n$, then $x_2 = \ldots = x_n = 0$, $x_1 = 1$ is a non-trivial solution, so suppose that $a_{i1} \neq 0$ for some i between 1 and n. In fact, without loss of generality, we may assume that $a_{11} \neq 0$.

Now subtract $(a_{i1}/a_{11}) \times$ equation 1 from equation i ($2 \leq i \leq n$) to produce the following equivalent system:

$$a_{11}x_1 + a_{12}x_2 + \ldots + a_{1n}x_n = 0$$
$$b_{22}x_2 + \ldots + b_{2n}x_n = 0$$
$$\cdots\cdots\cdots\cdots\cdots$$
$$b_{n2}x_2 + \ldots + b_{nn}x_n = 0 \quad \text{where } b_{ij} = a_{ij} - a_{i1}a_{1j}/a_{11}.$$

Let A' be the matrix of this new system, and let $\det B$ be the $(1,1)$-minor of A'.

Then $\det A = \det A'$ by Corollary 4.3.9

$\quad\quad\quad\quad = a_{11} \det B$ (expanding $\det A'$ by the first column).

But $\det A = 0$ and so $\det B = 0$ ($a_{11} \neq 0$, remember). It follows from the inductive hypothesis that the $(n-1) \times (n-1)$ system with matrix B has a non-trivial solution $x_2 = \alpha_2, \ldots, x_n = \alpha_n$. Substituting into the first equation of the original system we see that

$$x_1 = (1/a_{11})(-a_{12}\alpha_2 - \ldots - a_{1n}\alpha_n), x_2 = \alpha_2, \ldots, x_n = \alpha_n$$

is a non-trivial solution to $AX = O$. The result follows by induction.

COROLLARY 4.7.2 Let $A = [a_{ij}]_{m \times n}$. Then the system $AX = O$ has a non-trivial solution if $m < n$.

Proof Add $n - m$ equations of the form $0x_1 + \ldots + 0x_n = 0$. This produces an equivalent system whose matrix is $n \times n$ and singular. This system, and consequently the equivalent original one, each have a non-trivial solution by Theorem 4.7.1.

EXERCISES 4.7

1 Decide whether or not each of the following systems of equations has a non-trivial solution:

(a) $x + y = 0$
$x - y = 0$.

(b) $x + 3y - 2z = 0$
$-x + 4y + z = 0$
$5x - 6y - 7z = 0$.

(c) $x + 3y = 2z$
$2x + 4y + z = 0$
$x + 6y = 3z$.

2 For what value of a does the following system have a non-trivial solution?

$$ax - 3y + 4z = 0$$
$$x + y = 0$$
$$3x + ay + 2z = 0.$$

SOLUTIONS AND HINTS FOR EXERCISES

Exercises 4.1

1 (1) (the identity), (12), (13), (14), (23), (24), (34), (123), (132), (124), (142), (231), (213), (312), (321), (1234), (1243), (1324), (1342), (1423), (1432), (12)(34), (13)(24), (14)(23).

2 Even: (1), (123), (132); odd: (12), (13), (23).

3 Put $T_n = \{\sigma^{-1}: \sigma \in S_n\}$. Let $\sigma \in S_n$. Then σ is bijective, so $\sigma^{-1} = \psi$, say, exists in S_n. But $\sigma = \psi^{-1} \in T_n$, so $S_n \subseteq T_n$. Now let $\psi \in T_n$. Then $\psi = \sigma^{-1}$ where $\sigma \in S_n$, so $\psi \in S_n$. Thus $T_n \subseteq S_n$.

4 Suppose that $\sigma(i) \geq i$ for all $1 \leq i \leq n$. Then $\sigma(n) \geq n$, so $\sigma(n) = n$. But now $\sigma(n-1) = n - 1$ or n. We cannot have $\sigma(n-1) = n$ since $\sigma(n) = n$ and σ is injective. Thus, $\sigma(n-1) = n - 1$. A simple induction argument shows that $\sigma(i) = i$ for all $1 \leq i \leq n$. Hence $\sigma = (1)$.

5 Let $Q_n(x_2, \ldots, x_n) = \prod_{1 \leq j < i} (x_i - x_j)$

(so that $P_n(x_1, \ldots, x_n) = \prod_{k=2}^{n} (x_1 - x_k)Q_n(x_2, \ldots, x_n)$,
and put $\hat{\psi}(Q_n) = Q_n(x_{\psi(2)}, \ldots, x_{\psi(n)})$ for $\psi \in T_n$. Then
sign $(\psi) = 1$ if $\hat{\psi}(Q_n) = Q_n$, sign $(\psi) = -1$ if $\hat{\psi}(Q_n) = -Q_n$. Now

$$\text{sign}(\psi) = 1 \Leftrightarrow \hat{\psi}(Q_n) = Q_n \Leftrightarrow \hat{\sigma}(P_n) = P_n(x_{\sigma(1)}, \ldots, x_{\sigma(n)})$$
$$= P_n(x_1, x_{\psi(2)}, \ldots, x_{\psi(n)})$$
$$= \prod_{k=2}^{n}(x_1 - x_{\psi(k)})Q_n(x_{\psi(2)}, \ldots, x_{\psi(n)})$$
$$= \prod_{k=2}^{n}(x_1 - x_k)Q_n(x_2, \ldots, x_n) = P_n$$
$$\Leftrightarrow \text{sign}(\sigma) = 1.$$

Exercises 4.2

1 (a) -2; (b) 11; (c) -7; (d) 1.

2 The argument of exercise 4.1.4 is easily modified to show that if $\sigma \neq (1)$ then $\sigma(i) > i$ for some $1 \leq i \leq n$. Since $a_{ij} = 0$ if $i < j$, $a_{i\sigma(i)} = 0$. The rest of the argument is identical to Example 4.2.3.

3 Let $A = [a_{ij}]_{n \times n}$ be upper triangular. Then A is already in echelon form. The reduced echelon form can only have ones on the main diagonal if all of the diagonal elements are non-zero. The result now follows from examples 3.6.2 and 4.2.3 (In fact, the alert reader will probably have observed that this 'proof' is unsatisfactory. However, it serves the purpose here of making the reader think about some of the things we have been doing; we will prove the result 'properly' later.)

Exercises 4.3

1 (a) -6; (b) 0; (c) -100.

2 $\det(E_r(\alpha)) = \alpha$, $\det(E_{rs}(\alpha)) = 1$, $\det(E_{rs}) = -1$.

3 (a) If $E = E_r(\alpha)$, then $(\det E)(\det A) = \alpha \det A = \det(EA)$, by Theorem 4.3.3;
if $E = E_{rs}(\alpha)$, then $(\det E)(\det A) = \det A = \det(EA)$, by Corollary 4.3.9;
if $E = E_{rs}$, then $(\det E)(\det A) = -\det A = \det(EA)$, by Theorem 4.3.6.

(b) Use induction on m.

4 Choose $A = \begin{bmatrix} 1 & 0 \\ 0 & 0 \end{bmatrix}$, $B = \begin{bmatrix} 0 & 0 \\ 0 & 1 \end{bmatrix}$ for example.

Exercises 4.4

1 $\begin{vmatrix} 1-a & 1 & 1 \\ 2 & 1-a & 2 \\ 2 & -1 & 4-a \end{vmatrix} = \begin{vmatrix} 3-a & 0 & 3-a \\ 2 & 1-a & 2 \\ 2 & -1 & 4-a \end{vmatrix}$ $R_1 = r_1 + r_3$

$ = \begin{vmatrix} 3-a & 0 & 0 \\ 2 & 1-a & 0 \\ 2 & -1 & 2-a \end{vmatrix}$ $C_3 = c_3 - c_1$

$ = (3-a)(1-a)(2-a).$

Thus, the matrix is invertible if and only if $a = 1, 2$ or 3.

2 AB is invertible $\Leftrightarrow 0 \ne \det(AB) = (\det A)(\det B)$ (Theorem 4.4.2)

$\phantom{AB \text{ is invertible}} \Leftrightarrow \det A \ne 0$ and $\det B \ne 0$

$\phantom{AB \text{ is invertible}} \Leftrightarrow A$ and B are both invertible.

3 $\det(I_n - B^{-1}AB) = \det(B^{-1}(I_n - A)B) = \det(B^{-1})\det(I_n - A)\det B$

 (Theorem 4.4.2)

$ = \det(I_n - A)\det(B^{-1})\det B = \det(I_n - A)\det(B^{-1}B)$

 (Theorem 4.4.2)

$ = \det(I_n - A)\det(I_n) = \det(I_n - A).$

4 $A^T = -A \Rightarrow -\det A = (-1)^{2n+1}\det A = \det(-A)$ (Theorem 4.3.3)

$ = \det(A^T) = \det A$ (Corollary 4.3.2)

$ \Rightarrow 2\det A = 0 \Rightarrow \det A = 0 \Rightarrow A$ is singular.

Exercises 4.5

1 (a) 0;   (b) 42;   (c) 0.

2 $\begin{vmatrix} a-x & a-y & a-z \\ b-x & b-y & b-z \\ c-x & c-y & c-z \end{vmatrix} = \begin{vmatrix} a-b & a-b & a-b \\ b-c & b-c & b-c \\ c-x & c-y & c-z \end{vmatrix}$ $R_1 = r_1 - r_3$
 $R_2 = r_2 - r_3$

$ = \begin{vmatrix} a-b & 0 & 0 \\ b-c & 0 & 0 \\ c-x & x-y & x-z \end{vmatrix}$ $C_2 = c_2 - c_1$
 $C_3 = c_3 - c_1$

$ = (a-b)(b-c) \begin{vmatrix} 1 & 0 & 0 \\ 1 & 0 & 0 \\ c-x & x-y & x-z \end{vmatrix} = 0.$

3 Expanding by the first row gives

$$D_n = (1+x^2)D_{n-1} - x \begin{vmatrix} x & x & 0\ldots 0 & 0 \\ 0 & 1+x^2 & x\ldots 0 & 0 \\ \multicolumn{4}{c}{\cdots\cdots\cdots\cdots\cdots\cdots\cdots} \\ 0 & \ldots & x & 1+x^2 \end{vmatrix}$$

$$= (1+x^2)D_{n-1} - x \begin{vmatrix} x & 0 & 0\ldots 0 & 0 \\ 0 & 1+x^2 & x\ldots 0 & 0 \\ \multicolumn{4}{c}{\cdots\cdots\cdots\cdots\cdots\cdots\cdots} \\ 0 & \ldots & x & 1+x^2 \end{vmatrix} \quad C_2 = c_2 - c_1$$

$= (1+x^2)D_{n-1} - x^2 D_{n-2}$, expanding by the first row again.

Rearranging this equation gives the desired result. It is easy to check that this equation is valid for $n \geq 3$. Now

$$D_k - D_{k-1} = x^2(D_{k-1} - D_{k-2}) = x^4(D_{k-2} - D_{k-3})$$
$$= \ldots = x^{2k-4}(D_2 - D_1) = x^{2k}.$$

Thus $D_n = (D_n - D_{n-1}) + (D_{n-1} - D_{n-2}) + \ldots + (D_2 - D_1) + D_1$
$$= x^{2n} + x^{2n-2} + \ldots + x^4 + x^2 + 1 = (x^{2n+2} - 1)/(x^2 - 1).$$

4 $\begin{vmatrix} a & -b & -a & b \\ b & a & -b & -a \\ c & -d & c & -d \\ d & c & d & c \end{vmatrix} = \begin{vmatrix} a & -b & 0 & 0 \\ b & a & 0 & 0 \\ c & -d & c & -d \\ d & c & d & c \end{vmatrix} \quad \begin{matrix} C_3 = c_3 + c_1 \\ C_4 = c_4 + c_2 \end{matrix}$

$$= \begin{vmatrix} a & -b & 0 & 0 \\ b & a & 0 & 0 \\ 0 & 0 & c & -d \\ 0 & 0 & d & c \end{vmatrix} \quad \begin{matrix} C_1 = c_1 - c_3 \\ C_2 = c_2 - c_4 \end{matrix}$$

$$= 4(a^2 + b^2)(c^2 + d^2).$$

5 $\begin{vmatrix} a^2 & a & 1 \\ b^2 & b & 1 \\ c^2 & c & 1 \end{vmatrix} = \begin{vmatrix} a^2 - b^2 & a-b & 0 \\ b^2 - c^2 & b-c & 0 \\ c^2 & c & 1 \end{vmatrix} \quad \begin{matrix} R_1 = r_1 - r_2 \\ R_2 = r_2 - r_3 \end{matrix}$

$$= \begin{vmatrix} a^2 - b^2 & a-b \\ b^2 - c^2 & b-c \end{vmatrix} \quad \text{expanding by column 3}$$

$$= (a-b)(b-c)\begin{vmatrix} a+b & 1 \\ b+c & 1 \end{vmatrix} = (a-b)(b-c)(a-c).$$

Also $\begin{vmatrix} 1 & -2x & x^2 \\ 1 & -2y & y^2 \\ 1 & -2z & z^2 \end{vmatrix} = 2 \begin{vmatrix} x^2 & x & 1 \\ y^2 & y & 1 \\ z^2 & z & 1 \end{vmatrix} = 2(x-y)(y-z)(x-z).$

Thus
$\begin{vmatrix} (a-x)^2 & (a-y)^2 & (a-z)^2 \\ (b-x)^2 & (b-y)^2 & (b-z)^2 \\ (c-x)^2 & (c-y)^2 & (c-z)^2 \end{vmatrix} = \begin{vmatrix} a^2 & a & 1 \\ b^2 & b & 1 \\ c^2 & c & 1 \end{vmatrix} \begin{vmatrix} 1 & 1 & 1 \\ -2x & -2y & -2z \\ x^2 & y^2 & z^2 \end{vmatrix}$

$= 2(a-b)(b-c)(a-c)(x-y)(y-z)(x-z).$

6 The given determinant has value $a^2 + b^2 + c^2 + d^2$. Also

$\begin{vmatrix} a_1 + ib_1 & c_1 + id_1 \\ -(c_1 - id_1) & a_1 - ib_1 \end{vmatrix} \begin{vmatrix} a_2 + ib_2 & c_2 + id_2 \\ -(c_2 - id_2) & a_2 - ib_2 \end{vmatrix} = \begin{vmatrix} a + ib & c + id \\ -(c - id) & a - ib \end{vmatrix}$

where $a = a_1 a_2 - b_1 b_2 - c_1 c_2 - d_1 d_2$, $b = b_1 a_2 + a_1 b_2 + c_1 d_2 - c_2 d_1$, $c = a_1 c_2 - b_1 d_2 + c_1 a_2 + d_1 b_2$, $d = a_1 d_2 + b_1 c_2 - c_1 b_2 + d_1 a_2$. The result follows.

Exercises 4.6

1 $\text{adj } A = \begin{bmatrix} 3 & 2 & 6 \\ 1 & 1 & 2 \\ 2 & 2 & 5 \end{bmatrix}$, $\det A = 1$, so $A^{-1} = \text{adj } A$.

2 (a) If A is non-singular, $\text{adj } A = (\det A) A^{-1}$ (Corollary 4.6.3). Hence

$(\text{adj } A^{-1})(\text{adj } A)^{-1} = (\det A^{-1})(A^{-1})^{-1}(\det A) A^{-1} = \det(A^{-1} A) A A^{-1} = I_n.$

Therefore $\text{adj } A^{-1} = (\text{adj } A)^{-1}.$

(b) Here

$\text{adj}(AB) = \det(AB)(AB)^{-1} = (\det A)(\det B) B^{-1} A^{-1}$

$= ((\det B) B^{-1})((\det A) A^{-1})$

$= (\text{adj } B)(\text{adj } A).$

(c) We have $A \text{ adj } A = (\det A) I_n$, by Theorem 4.6.2. Taking determinants of both sides of this equation,

$(\det A)(\det(\text{adj } A)) = (\det A)^n$, whence $\det(\text{adj } A) = (\det A)^{n-1}.$

Thus $\text{adj}(\text{adj } A) = \det(\text{adj } A)(\text{adj } A)^{-1}$

$= (\det A)^{n-1}(\det A)^{-1} A$ (Theorem 4.6.2)

$= (\det A)^{n-2} A.$

Hence $\quad \text{adj}(\text{adj}(\text{adj } A)) = \det(\text{adj}(\text{adj } A))(\text{adj}(\text{adj } A))^{-1}$
$$= (\det A)^{n(n-2)}(\det A)(\det A)^{2-n} A^{-1}$$
$$= (\det A)^{n^2 - 3n + 3} A^{-1}.$$

3 Suppose first that A is singular. Then $\det A = 0$, and so $A(\text{adj } A) = O$. Suppose that $\text{adj } A$ is non-singular, so there is a B such that $(\text{adj } A)B = I_n$, and hence
$$A = AI_n = A(\text{adj } A)B = OB = O.$$

Clearly then $\text{adj } A = O$, which is singular. This contradiction gives the result.

Also, if A is non-singular, then $\det A \neq O$. But now
$(\det A)(\det(\text{adj } A)) = (\det A)^n$ (as in 2(c) above), and therefore
$\det(\text{adj } A) = (\det A)^{n-1}.$

If A is singular then $\det A = 0$, $\det(\text{adj } A) = 0$, and so the equation is still valid.

4 Clearly, $0 = \omega^3 - 1 = (\omega - 1)(\omega^2 + \omega + 1)$. Since $\omega - 1 \neq 0$,
$\omega^2 + \omega + 1 = 0$. It is then easy to check that $A^{-1} = \frac{1}{3}\begin{bmatrix} 1 & 1 & 1 \\ 1 & \omega^2 & \omega \\ 1 & \omega & \omega^2 \end{bmatrix}$.

Exercises 4.7

1 Writing each of these in the form $AX = 0$, we have

(a) $\det A = -2$, so there is only the trivial solution;
(b) $\det A = 0$, so there is a non-trivial solution;
(c) $\det A = -13$, so there is only the trivial solution.

2 $\begin{vmatrix} a & -3 & 4 \\ 1 & 1 & 0 \\ 3 & a & 2 \end{vmatrix} = \begin{vmatrix} a & -3-a & 4 \\ 1 & 0 & 0 \\ 3 & a-3 & 2 \end{vmatrix} \quad C_2 = c_2 - c_1$

$$= -1 \begin{vmatrix} -3-a & 4 \\ a-3 & 2 \end{vmatrix} = \begin{vmatrix} a+3 & 4 \\ 3-a & 2 \end{vmatrix} = 6a - 6 = 0 \Leftrightarrow a = 1.$$

Thus, there is a non-trivial solution if and only if $a = 1$.

5 VECTOR SPACES

5.1 INTRODUCTION

Let us start reviewing the situation we studied in Chapter 1. We were concerned with two sets: a set V of vectors and a set F of scalars. We defined a means of adding vectors and of multiplying vectors by scalars, and found that these two operations satisfied the following axioms, or laws:

V1 $(\mathbf{a} + \mathbf{b}) + \mathbf{c} = \mathbf{a} + (\mathbf{b} + \mathbf{c})$ for all $\mathbf{a}, \mathbf{b}, \mathbf{c} \in V$;

V2 there is a vector $\mathbf{0} \in V$ with the property that
$$\mathbf{a} + \mathbf{0} = \mathbf{0} + \mathbf{a} = \mathbf{a} \qquad \text{for all } \mathbf{a} \in V;$$

V3 for each $\mathbf{a} \in V$ there is a corresponding vector $-\mathbf{a} \in V$ such that
$$\mathbf{a} + (-\mathbf{a}) = (-\mathbf{a}) + \mathbf{a} = \mathbf{0};$$

V4 $\mathbf{a} + \mathbf{b} = \mathbf{b} + \mathbf{a}$ for all $\mathbf{a}, \mathbf{b} \in V$;

V5 $\alpha(\mathbf{a} + \mathbf{b}) = \alpha\mathbf{a} + \alpha\mathbf{b}$ for all $\alpha \in F$, and all $\mathbf{a}, \mathbf{b} \in V$;

V6 $(\alpha + \beta)\mathbf{a} = \alpha\mathbf{a} + \beta\mathbf{a}$ for all $\alpha, \beta \in F$, and all $\mathbf{a} \in V$;

V7 $(\alpha\beta)\mathbf{a} = \alpha(\beta\mathbf{a})$ for all $\alpha, \beta \in F$, and all $\mathbf{a} \in V$;

V8 $1\mathbf{a} = \mathbf{a}$ for all $\mathbf{a} \in V$;

V9 $0\mathbf{a} = \mathbf{0}$ for all $\mathbf{a} \in V$.

In Chapter 1 the set V was either $\mathbb{R}^3$ or $\mathbb{R}^2$, and F was the set of real numbers. This same collection of properties was encountered again in Chapter 2 in a different context. Here the set V could be the set of all $m \times n$ real matrices and F the set of real numbers.

The idea we want to exploit here is that any results which could be deduced directly from these axioms, V1 to V9, would apply equally to $\mathbb{R}^3$, to $\mathbb{R}^2$, to sets of matrices and to any other situation in which they are satisfied. Such situations, as we will see, abound. So we could study all these systems simultaneously. This economy of effort is in itself appealing, but there are also other advantages in this approach. For example, it helps to place our study on a firm foundation. If we treat the axioms as 'the rules of the game', then we have made precise what our assumptions are. We can also gain inspiration from one area and apply it in another; for example, the geometry implicit in $\mathbb{R}^3$ suggests concepts of utility in areas such as polynomial theory where their significance might not otherwise have been perceived. Moreover, it can sometimes be easier to prove results in this abstract setting: in a sense, all of the superfluous information has been trimmed away and we are no longer prevented from seeing the wood for the trees.

Of course, there is a price to be paid: the resulting theory has a higher level of abstraction. You should not be deterred by this, however, as the process is one with which you have coped before, from your earliest experiences with mathematics. When you first started school you were introduced to the abstract concept of number. You were shown many sets of three objects—three balls, three dogs, three pencils—and, gradually, you learned to recognise the property that they had in common, namely, their 'threeness', and gave a name to it. Here the procedure is similar; we are taking a series of situations, recognising something that they have in common (that they satisfy this set of axioms) and giving a name to it.

So what is the name? A *vector space* consists of two sets, V and F, the elements of which, for convenience, we will refer to as *vectors* and *scalars*, respectively. (Though, by vectors we do not mean that they must be elements of $\mathbb{R}^2$ or of $\mathbb{R}^3$. We make no assumptions other than those we specify!) There must be a rule for combining vectors $\mathbf{v}_1, \mathbf{v}_2 \in V$ to give a vector $\mathbf{v}_1 + \mathbf{v}_2 \in V$, and a rule for combining any vector $\mathbf{v} \in V$ and any scalar $\alpha \in F$ to give a vector $\alpha \mathbf{v} \in V$. Furthermore, axioms V1 to V9 must be satisfied.

Actually, we *have* glossed over one point. In axioms V6 and V7 we speak of $\alpha + \beta$ and $\alpha\beta$, and in V8 and V9 it is implicit that $0, 1 \in F$, so that F is not just any old set; it must be another abstract construct known as a *field*. However, we do not wish to spell out the precise axioms which define a field if you are not already familiar with them; let us concentrate on absorbing one set of axioms at a time. Suffice it to say that a field is a set of 'numbers' in which we can 'add', 'subtract', 'multiply' and 'divide', and in which all of the usual laws of arithmetic are satisfied. Examples of fields are $\mathbb{Q}, \mathbb{R}$ and $\mathbb{C}$ ($\mathbb{Z}$ is not a field because we cannot divide in $\mathbb{Z}$). In all of the subsequent development nothing will be lost if you think of $\mathbb{Q}, \mathbb{R}$ or $\mathbb{C}$ whenever we use F.

We speak of the *vector space V over the field F*.

Examples 5.1

1. $\mathbb{R}^2$ is a vector space over $\mathbb{R}$.
2. $\mathbb{R}^3$ is a vector space over $\mathbb{R}$.

3. Let $M_{m,n}(\mathbb{R})$ $[M_{m,n}(\mathbb{C})]$ be the set of all real (complex) $m \times n$ matrices. Then $M_{m,n}(\mathbb{R})$ is a vector space over $\mathbb{R}$, and $M_{m,n}(\mathbb{C})$ is a vector space over $\mathbb{C}$ *and* over $\mathbb{R}$. ($M_{m,n}(\mathbb{R})$ is *not* a vector space over $\mathbb{C}$, as multiplication of a real matrix by a complex number does not necessarily produce a real matrix.)

4. Let $\mathbb{R}^n$ denote the set of all ordered n-tuples of real numbers, so that

$$\mathbb{R}^n = \{(x_1, \ldots, x_n): x_1, \ldots, x_n \in \mathbb{R}\}.$$

Define an addition and a scalar multiplication on $\mathbb{R}^n$ by

$$(x_1, \ldots, x_n) + (y_1, \ldots, y_n) = (x_1 + y_1, \ldots, x_n + y_n),$$

$$\alpha(x_1, \ldots, x_n) = (\alpha x_1, \ldots, \alpha x_n) \quad \text{for all } \alpha, x_1, \ldots, x_n, y_1, \ldots, y_n \in \mathbb{R}.$$

Then $\mathbb{R}^n$ has a zero element $(0, \ldots, 0)$, each element $(x_1, \ldots, x_n) \in \mathbb{R}^n$ has an additive inverse $(-x_1, \ldots, -x_n)$, and it is easy to check that all of the axioms V1 to V9 are satisfied. Hence, $\mathbb{R}^n$ is a vector space over $\mathbb{R}$.

5. $\mathbb{C}$ is a vector space over $\mathbb{R}$. We have a means of adding complex numbers, and of multiplying any complex number by a real number, namely

$$(a + bi) + (c + di) = (a + c) + (b + d)i,$$

$$\alpha(a + bi) = \alpha a + \alpha b i \quad \text{for all } \alpha, a, b, c, d \in \mathbb{R}.$$

There is a zero element, $0 + 0i$, each complex number $a + bi$ has an additive inverse, $-a - bi$, and it is straightforward to check (and probably well-known to you already) that axioms V1 to V9 are satisfied.

6. Let P_n be the set of real polynomials of degree less than or equal to n; that is,

$$P_n = \{a_0 + a_1 x + \ldots + a_n x^n : a_0, a_1, \ldots, a_n \in \mathbb{R}\}.$$

We can define an addition and a scalar multiplication by

$$(a_0 + a_1 x + \ldots + a_n x^n) + (b_0 + b_1 x + \ldots + b_n x^n)$$
$$= (a_0 + b_0) + (a_1 + b_1) x + \ldots + (a_n + b_n) x^n,$$

and $\quad \alpha(a_0 + a_1 x + \ldots + a_n x^n) = \alpha a_0 + \alpha a_1 x + \ldots + \alpha a_n x^n$

for all $\alpha, a_i, b_i \in \mathbb{R}$.

The zero polynomial is $0 + 0x + \ldots + 0x^n$ and the additive inverse of $a_0 + a_1 + \ldots + a_n x^n$ is $-a_0 - a_1 x - \ldots - a_n x^n$. Then rugged determination is all that is required to check that P_n is a vector space over $\mathbb{R}$!

Having laid down this set of axioms, these are now the rules by which we must abide. We must be careful not to make any assumptions other than these, or the results obtained may not be true for the examples to which we wish to apply them. As a simple illustration of working from the axioms we will close the section by collating some elementary consequences of the axioms, some of which we will prove, while we will leave others as exercises.

LEMMA 5.1.1 Let V be a vector space over a field F. Then

(a) the zero element $\mathbf{0}$ is unique;

(b) for each $\mathbf{v} \in V$, the element $-\mathbf{v}$ is uniquely defined;

(c) $\alpha \mathbf{0} = \mathbf{0}$ for all $\alpha \in F$;

(d) $(-\alpha)\mathbf{v} = -(\alpha \mathbf{v})$ for all $\alpha \in F$, and all $\mathbf{v} \in V$;

(e) $-(-\mathbf{v}) = \mathbf{v}$ for all $\mathbf{v} \in V$;

(f) $\alpha(-\mathbf{v}) = -(\alpha \mathbf{v})$ for all $\alpha \in F$, and all $\mathbf{v} \in V$;

(g) $(-\alpha)(-\mathbf{v}) = \alpha \mathbf{v}$ for all $\alpha \in F$, and all $\mathbf{v} \in V$.

Proof

(a) Suppose that $\mathbf{o}$ is also a zero element for V. Then $\mathbf{0} + \mathbf{a} = \mathbf{a}$ for all $\mathbf{a} \in V$ by V2, and $\mathbf{b} + \mathbf{o} = \mathbf{b}$ for all $\mathbf{b} \in V$, also by V2. In particular, putting $\mathbf{a} = \mathbf{o}$ and $\mathbf{b} = \mathbf{0}$ we see that $\mathbf{0} + \mathbf{o} = \mathbf{o}$ and that $\mathbf{0} + \mathbf{o} = \mathbf{0}$. Hence $\mathbf{o} = \mathbf{0}$.

(c) Since $\alpha \mathbf{0} \in V$, ther is a vector $-(\alpha \mathbf{0}) \in V$, by V3. Thus

$$\begin{aligned}\mathbf{0} &= \alpha \mathbf{0} + -(\alpha \mathbf{0}) && \text{by V3} \\ &= \alpha(\mathbf{0} + \mathbf{0}) + -(\alpha \mathbf{0}) && \text{by V2} \\ &= (\alpha \mathbf{0} + \alpha \mathbf{0}) + -(\alpha \mathbf{0}) && \text{by V5} \\ &= \alpha \mathbf{0} + (\alpha \mathbf{0} + -(\alpha \mathbf{0})) && \text{by V1} \\ &= \alpha \mathbf{0} + \mathbf{0} && \text{by V3} \\ &= \alpha \mathbf{0} && \text{by V2.}\end{aligned}$$

(d) Here we have to show that $(-\alpha)\mathbf{v}$ is the additive inverse of $\alpha \mathbf{v}$. To establish this we show that it has the property claimed for the additive inverse in V3; namely that when added to $\alpha \mathbf{v}$ we get $\mathbf{0}$. Now

$$\begin{aligned}(-\alpha)\mathbf{v} + \alpha \mathbf{v} &= (-\alpha + \alpha)\mathbf{v} && \text{by V6} \\ &= 0\mathbf{v} && \text{because } F \text{ is a field} \\ &= \mathbf{0} && \text{by V9}\end{aligned}$$

Similarly, $\alpha \mathbf{v} + (-\alpha)\mathbf{v} = \mathbf{0}$.

The remaining parts of this lemma are set as exercise 5.1.5.

EXERCISES 5.1

1 Let $\mathscr{S}$ be the set of all infinite sequences of real numbers, so that

$$\mathscr{S} = \{(a_1, a_2, a_3, \ldots): a_i \in \mathbb{R} \text{ for } i \geq 1\}.$$

Define

$$(a_1, a_2, a_3, \ldots) + (b_1, b_2, b_3, \ldots) = (a_1 + b_1, a_2 + b_2, a_3 + b_3, \ldots)$$

$$\alpha(a_1, a_2, a_3, \ldots) = (\alpha a_1, \alpha a_2, \alpha a_3, \ldots) \quad \text{for all } \alpha, a_i, b_i \in \mathbb{R}.$$

Check that $\mathscr{S}$ together with these operations is a vector space over $\mathbb{R}$.

2 Let $\mathbb{Q}(\sqrt{2}) = \{a + b\sqrt{2} : a, b \in \mathbb{Q}\}$ and define

$$(a + b\sqrt{2}) + (c + d\sqrt{2}) = (a + c) + (b + d)\sqrt{2}$$

$$\alpha(a + b\sqrt{2}) = \alpha a + \alpha b \sqrt{2} \quad \text{for all } \alpha, a, b, c, d \in \mathbb{Q}.$$

Check that $\mathbb{Q}(\sqrt{2})$ with these operations forms a vector space over $\mathbb{Q}$.

3 Let Map($\mathbb{R}, \mathbb{R}$) denote the set of all mappings from $\mathbb{R}$ into itself. An addition and a scalar multiplication can be defined on Map($\mathbb{R}, \mathbb{R}$) by

$$(f + g)(x) = f(x) + g(x)$$

$$(\alpha f)(x) = \alpha f(x) \quad \text{for all } \alpha, x \in \mathbb{R}, \quad \text{and all } f, g \in \text{Map}(\mathbb{R}, \mathbb{R}).$$

Show that Map($\mathbb{R}, \mathbb{R}$) together with these operations forms a vector space over $\mathbb{R}$.

4 Show that axiom V9 is, in fact, redundant: it can be deduced from the other axioms.

5 Prove Lemma 5.1.1 parts (b), (e), (f) and (g).

5.2 SUBSPACES AND SPANNING SEQUENCES

It is possible for one vector space to sit inside another. For example, we can think of $\mathbb{R}^2$ as sitting inside $\mathbb{R}^3$, of P_{n-1} as being inside P_n, or of $\mathbb{R}$ as lying in $\mathbb{C}$. We formalise this idea as shown below.

If V is a vector space over the field F, then any subset of V which also forms a vector space over F, with the same definitions of vector addition and scalar multiplication, is called a *subspace* of V. In order to show that a subset W of V is a subspace it is not necessary to check all of the vector space axioms: many of them follow automatically from the fact that W is lying *inside* a vector space, namely V. We will prove this in the next lemma.

LEMMA 5.2.1 Let V be a vector space over the field F, and let W be a subset of V. Then W is a subspace of V if and only if the following conditions hold:

S1 W is non-empty;

S2 $\mathbf{w}_1, \mathbf{w}_2 \in W \Rightarrow \mathbf{w}_1 + \mathbf{w}_2 \in W$;

S3 $\alpha \in F, \mathbf{w} \in W \Rightarrow \alpha \mathbf{w} \in W$.

Proof Suppose first that S1, S2 and S3 hold. Conditions S1 and S2 tell us that combining two vectors, or a vector and a scalar, under the vector addition and scalar multiplication of V does produce an element of W. We say that W is *closed* under addition and under scalar multiplication.

For each $\mathbf{w} \in W$, we have $(-1)\mathbf{w} \in W$, by S3, and $(-1)\mathbf{w} = -\mathbf{w}$, by Lemma 5.1.1(d). Now there *is* an element $\mathbf{w} \in W$, by S1, and $\mathbf{0} = \mathbf{w} + (-\mathbf{w}) \in W$ by S2, so it is clear that V2 and V3 are satisfied. All of the other axioms are inherited from V: they hold for all elements in V, so, in particular, they hold for all elements in W.

The converse is easily seen to be true.

Examples 5.2.1

1. Every vector space V has two trivial subspaces, namely the zero subspace, $\{\mathbf{0}\}$, and V itself.

2. Let $V = \mathbb{R}^2$ and let $W = \{(x, y) \in \mathbb{R}^2 : x + y = 0\}$. Then W is a subspace of V. For, $(0, 0) \in W$, so S1 is satisfied. If $(x_1, y_1), (x_2, y_2) \in W$, then $x_1 + y_1 = x_2 + y_2 = 0$.
Thus $(x_1, y_1) + (x_2, y_2) = (x_1 + x_2, y_1 + y_2) \in W$ because $(x_1 + x_2) + (y_1 + y_2) = (x_1 + y_1) + (x_2 + y_2) = 0$ and S2 holds.
Also, if $(x, y) \in W$, then $x + y = 0$, and so $\alpha x + \alpha y = 0$ for all $\alpha \in \mathbb{R}$. Hence $\alpha(x, y) = (\alpha x, \alpha y) \in W$ and S3 is satisfied.

3. Let $V = \mathbb{R}^3$ and let $W = \{(x, y, z) \in \mathbb{R}^3 : 2x + 3y - z = 0\}$. Then W is a subspace of V. The proof is similar to that given in example 2 above.

4. Let $V = P_2$ and let $W = \{f(x) \in P_2 : f(x) \text{ has degree } 2\}$. Then W is *not* a subspace of V. For example, $x^2 + x + 1, -x^2 + x + 1 \in W$, but their sum is $2x + 2 \notin W$, so S2 is not satisfied.

5. Let V be any vector space over the field F, and let $\mathbf{v}_1, \ldots, \mathbf{v}_n \in V$. Put $W = \{\alpha_1 \mathbf{v}_1 + \ldots + \alpha_n \mathbf{v}_n : \alpha_1, \ldots, \alpha_n \in F\}$. Then $\mathbf{v}_1 \in W$ (choose $\alpha_1 = 1$, $\alpha_i = 0$ if $i \neq 1$), so S1 is satisfied. Suppose that $\mathbf{w}_1, \mathbf{w}_2 \in W$. Then $\mathbf{w}_1 = \alpha_1 \mathbf{v}_1 + \ldots + \alpha_n \mathbf{v}_n$, $\mathbf{w}_2 = \beta_1 \mathbf{v}_1 + \ldots + \beta_n \mathbf{v}_n$ where $\alpha_1, \ldots, \alpha_n, \beta_1, \ldots, \beta_n \in F$, and
$$\mathbf{w}_1 + \mathbf{w}_2 = (\alpha_1 + \beta_1)\mathbf{v}_1 + \ldots + (\alpha_n + \beta_n)\mathbf{v}_n \in W,$$
so S2 holds. Also, $\alpha \mathbf{w}_1 = \alpha \alpha_1 \mathbf{v}_1 + \ldots + \alpha \alpha_n \mathbf{v}_n \in W$, which confirms S3. Hence W is a subspace of V.

Example 5 above is an important one. In this situation we say that $\mathbf{v}_1, \ldots, \mathbf{v}_n$ *span* W, and we call W the *subspace spanned by* $\mathbf{v}_1, \ldots, \mathbf{v}_n$; we denote it by

$\langle \mathbf{v}_1, \ldots, \mathbf{v}_n \rangle$. Adopting the terminology of section 1.2, we say that every element of $\langle \mathbf{v}_1, \ldots, \mathbf{v}_n \rangle$ is a *linear combination* of $\mathbf{v}_1, \ldots, \mathbf{v}_n$.

Examples 5.2.2

1. Let $V = \mathbb{R}^2$, and let $\mathbf{v} \in V$. Then $\langle \mathbf{v} \rangle = \{\alpha \mathbf{v}: \alpha \in \mathbb{R}\}$ which is the origin if $\mathbf{v} = \mathbf{0}$, and is represented by a line through the origin if $\mathbf{v} \neq \mathbf{0}$. Let $\mathbf{v}_1, \mathbf{v}_2 \in V$ with $\mathbf{v}_1, \mathbf{v}_2$ lying in different directions. It follows from section 1.2 that every vector in the plane can be written as a linear combination of $\mathbf{v}_1, \mathbf{v}_2$, so that $\langle \mathbf{v}_1, \mathbf{v}_2 \rangle = \mathbb{R}^2$.

2. Let $V = P_n$ and let $1, x, \ldots, x^n \in P_n$. Then
$$\langle 1, x, \ldots, x^n \rangle = \{a_0 + a_1 x + \ldots + a_n x^n : a_0, a_1, \ldots, a_n \in \mathbb{R}\} = P_n,$$
so that $1, x, \ldots, x^n$ span P_n.

3. Let $V = \mathbb{C}$, considered as a vector space over $\mathbb{R}$, and let $1, i \in V$. Then
$$\langle 1, i \rangle = \{a1 + bi : a, b \in \mathbb{R}\} = \mathbb{C},$$
so $1, i$ span $\mathbb{C}$. Also
$$\langle 1 \rangle = \{a1 : a \in \mathbb{R}\} = \mathbb{R}.$$

4. Let $V = \mathbb{R}^4$, and let $W = \{(x, y, z, w) \in \mathbb{R}^4 : x + y = z - w = 0\}$. Then W is a subspace of V; the proof is similar to that given in example 5.2.1.2 above. Let $\mathbf{v} \in W$. Then $\mathbf{v} = (x, y, z, w)$ where $x + y = z - w = 0$. Now $y = -x, w = z$, so
$$\mathbf{v} = (x, -x, z, z) = (x, -x, 0, 0) + (0, 0, z, z) = x(1, -1, 0, 0) + z(0, 0, 1, 1).$$
Hence
$$W = \langle (1, -1, 0, 0), (0, 0, 1, 1) \rangle.$$

EXERCISES 5.2

1 Show that the vectors $\mathbf{v}_1 = (1, 0, 0, \ldots, 0)$, $\mathbf{v}_2 = (0, 1, 0, \ldots, 0)$, $\ldots$, $\mathbf{v}_n = (0, 0, 0, \ldots, 1)$ span $\mathbb{R}^n$.

2 Let $V = \mathbb{R}^3$, and suppose that $\mathbf{v}_1, \mathbf{v}_2, \mathbf{v}_3 \in \mathbb{R}^3$ are such that they do not all lie in the same plane. Describe geometrically $\langle \mathbf{v}_1 \rangle$, $\langle \mathbf{v}_1, \mathbf{v}_2 \rangle$ and $\langle \mathbf{v}_1, \mathbf{v}_2, \mathbf{v}_3 \rangle$.

3 Which of the following sets are subspaces of $\mathbb{R}^3$? In each case justify your answer.

(a) $\{(x, y, z) \in \mathbb{R}^3 : x + y + z = 1\}$.
(b) $\{(x, y, z) \in \mathbb{R}^3 : x - y = z\}$.
(c) $\{(x, y, z) \in \mathbb{R}^3 : x^2 + y^2 + z^2 = 0\}$.
(d) $\{(x, y, z) \in \mathbb{R}^3 : x^2 + y^2 + z^2 = 1\}$.
(e) $\{(x, y, z) \in \mathbb{R}^3 : x + y^2 = 0\}$.

4 Which of the following are subspaces of $V = \text{Map}(\mathbb{R}, \mathbb{R})$? Again justify your answers.

(a) $\{f \in V: f(1) = 1\}$.
(b) $\{f \in V: f(1) = 0\}$.
(c) $\{f \in V: f(x) = f(0) \text{ for all } x \in \mathbb{R}\}$.
(d) $\{f \in V: f(x) = f(1) \text{ for all } x \in \mathbb{R}\}$.
(e) $\{f \in V: f(x) \geq 0 \text{ for all } x \in \mathbb{R}\}$.

5 Find a spanning sequence for $\{f(x) \in P_n: f(0) = 0\}$.

6 Show that the sequence $\begin{bmatrix} 1 & 0 \\ 0 & 0 \end{bmatrix}, \begin{bmatrix} 0 & 1 \\ 0 & 0 \end{bmatrix}, \begin{bmatrix} 0 & 0 \\ 1 & 0 \end{bmatrix}, \begin{bmatrix} 0 & 0 \\ 0 & 1 \end{bmatrix}$ is a spanning sequence for $M_{2,2}(\mathbb{R})$ (see example 5.1.3).

7 Show that the matrices $\begin{bmatrix} 1 & 0 \\ 0 & 0 \end{bmatrix}, \begin{bmatrix} 1 & 1 \\ 0 & 0 \end{bmatrix}, \begin{bmatrix} 1 & 1 \\ 1 & 0 \end{bmatrix}, \begin{bmatrix} 1 & 1 \\ 1 & 1 \end{bmatrix}$ span $M_{2,2}(\mathbb{R})$.

8 Let U, W be subspaces of the vector space V. Prove that (a) $U \cap W$ and (b) $U + W = \{u + w: u \in U \text{ and } w \in W\}$ are subspaces of V. Prove that (c) $U \cup W$ is a subspace of V if and only if $U \subset W$ or $W \subset U$.

5.3 LINEAR INDEPENDENCE AND BASES

Here we want to adopt into our more general setting the ideas expounded in section 1.2 for $\mathbb{R}^2$ and $\mathbb{R}^3$. In fact, the ideas require very little translation. Throughout the section V will denote a vector space over the field F.

The sequence $v_1, \ldots, v_n$ of vectors in V is *linearly dependent* if one of them can be written as a linear combination of the rest; otherwise it is said to be *linearly independent*. Then Lemma 1.2.1 is also valid; in fact, precisely the same proof can be applied.

LEMMA 5.3.1

(a) $v_1, \ldots, v_n \in V$ are linearly dependent if and only if there are $\alpha_1, \ldots, \alpha_n \in F$, not all zero, such that $\alpha_1 v_1 + \ldots + \alpha_n v_n = \mathbf{0}$;
(b) $v_1, \ldots, v_n \in V$ are linearly independent if and only if
$$(\alpha_1 v_1 + \ldots + \alpha_n v_n = \mathbf{0} \Rightarrow \alpha_1 = \ldots = \alpha_n = 0).$$

Examples 5.3.1

1. The vectors $v_1 = (1, 0, \ldots, 0)$, $v_2 = (0, 1, 0, \ldots, 0)$, $\ldots$, $v_n = (0, \ldots, 0, 1)$ are

linearly independent elements of $\mathbb{R}^n$. For

$$\alpha_1 \mathbf{v}_1 + \alpha_2 \mathbf{v}_2 + \ldots + \alpha_n \mathbf{v}_n = (\alpha_1, \alpha_2, \ldots, \alpha_n) = (0, 0, \ldots, 0)$$
$$\Rightarrow \alpha_1 = \alpha_2 = \ldots = \alpha_n = 0,$$

and so linear independence follows from Lemma 5.3.1(b).

2. The vectors $(1, 2, -1, 1)$, $(1, 2, 1, 3)$, $(0, 0, -1, -1)$ are linearly dependent in $\mathbb{R}^4$. For

$$(1, 2, -1, 1) = (1, 2, 1, 3) + 2(0, 0, -1, -1).$$

3. The elements $1, x, \ldots, x^n$ are linearly independent in P_n. For

$$\alpha_0 1 + \alpha_1 x + \ldots + \alpha_n x^n = 0 + 0x + \ldots + 0x^n \Rightarrow \alpha_0 = \alpha_1 = \ldots = \alpha_n = 0.$$

4. The complex numbers $1, i$ are linearly independent in $\mathbb{C}$ (over $\mathbb{R}$). For

$$a, b \in \mathbb{R}, a1 + bi = 01 + 0i \Rightarrow a = b = 0.$$

The idea of a coordinate system which results from that of a basis is an important construct in $\mathbb{R}^2$ and $\mathbb{R}^3$. How can we imitate it in our new situation? The aspects of a basis which combine to give it its utility are that the vectors in it are linearly independent, which means essentially that it contains no redundant elements, and that every vector in the space can be written as a linear combination of the elements in it. We define, therefore, a (finite) sequence $\mathbf{v}_1, \ldots, \mathbf{v}_n$ of vectors in V to be a *basis* for V if it is a linearly independent sequence which spans V.

Examples 5.3.2

1. The vectors $\mathbf{v}_1 = (1, 0, \ldots, 0)$, $\mathbf{v}_2 = (0, 1, 0, \ldots, 0)$, $\ldots$, $\mathbf{v}_n = (0, \ldots, 0, 1)$ form a basis for $\mathbb{R}^n$ (see exercise 5.2.1 and example 5.3.1.1). We call it the *standard basis* for $\mathbb{R}^n$.

2. The elements $1, x, \ldots, x^n$ form a basis for P_n (see example 5.2.2.2 and example 5.3.1.3).

3. The complex numbers $1, i$ form a basis for $\mathbb{C}$, considered as a vector space over $\mathbb{R}$ (see example 5.2.2.3 and example 5.3.1.4).

4. The elements $1, 1 + x, 1 + x + x^2, \ldots, 1 + x + \ldots + x^n$ also form a basis for P_n. To establish this we need to check linear independence and spanning.

    Linear independence   Suppose that

    $$a_0 1 + a_1(1 + x) + \ldots + a_n(1 + x + \ldots + x^n) = 0.$$

VECTOR SPACES

Then
$$(a_0 + a_1 + \ldots + a_n) + (a_1 + \ldots + a_n)x + \ldots + a_n x^n = 0,$$
and so
$$a_0 + a_1 + \ldots + a_n = 0, \ a_1 + \ldots + a_n = 0, \ \ldots, \ a_n = 0.$$
It is easy to see that this system of homogeneous linear equations has only the trivial solution $a_0 = a_1 = \ldots = a_n = 0$.

Spanning Let $a_0 + a_1 x + \ldots + a_n x^n \in P_n$.
Now
$$a_0 + a_1 x + \ldots + a_n x^n = \alpha_0 1 + \alpha_1 (1+x) + \ldots + \alpha_n (1 + x + \ldots + x^n)$$
$$\Leftrightarrow a_0 = \alpha_0 + \alpha_1 + \ldots + \alpha_n, \ a_1 = \alpha_1 + \ldots + \alpha_n, \ \ldots, \ a_n = \alpha_n$$
$$\Leftrightarrow \alpha_n = a_n, \ \alpha_{n-1} = a_{n-1} - a_n, \ \ldots, \ \alpha_1 = a_1 - a_2, \ \alpha_0 = a_0 - a_1.$$

Thus
$$a_0 + a_1 x + \ldots + a_n x^n = (a_0 - a_1)1 + (a_1 - a_2)(1 + x)$$
$$+ \ldots + a_n(1 + x + \ldots + x^n),$$

and so every element of P_n can be written as a linear combination of the given polynomials.

LEMMA 5.3.2 Let $v_1, \ldots, v_n$ be a basis for V. Then every element of V can be written as a *unique* linear combination of $v_1, \ldots, v_n$.

Proof The proof is similar to that given in section 1.2 and will be set as an exercise.

EXERCISES 5.3

1 Show that the matrices $\begin{bmatrix} 1 & 0 \\ 0 & 0 \end{bmatrix}, \begin{bmatrix} 0 & 1 \\ 0 & 0 \end{bmatrix}, \begin{bmatrix} 0 & 0 \\ 1 & 0 \end{bmatrix}, \begin{bmatrix} 0 & 0 \\ 0 & 1 \end{bmatrix}$ are linearly independent elements of $M_{2,2}(\mathbb{R})$, and hence that they form a basis for $M_{2,2}(\mathbb{R})$.

2 Check whether the following are linearly dependent or independent:
   (a) $(1, 0, -1, 1), (0, 1, -3, 2), (-1, 2, 0, 1), (0, 4, 0, -1)$ in $\mathbb{R}^4$;
   (b) $1 - x, \ 1 + x, \ 1 - x + 2x^2$ in P_2;
   (c) $1 + i, \ 1 - i, \ 2 + 3i$ in $\mathbb{C}$ (as a vector space over $\mathbb{R}$);

(d) $\begin{bmatrix} 1 & -1 \\ -1 & 1 \end{bmatrix}, \begin{bmatrix} -1 & 1 \\ 1 & -1 \end{bmatrix}, \begin{bmatrix} 1 & 2 \\ 3 & 4 \end{bmatrix}, \begin{bmatrix} 0 & -3 \\ -4 & -3 \end{bmatrix}$ in $M_{2,2}(\mathbb{R})$.

3 Show that each of the following sequences form a basis for the given vector space:

(a) $(1, -2, -1), (-1, 0, 3), (0, 2, -1)$ for $\mathbb{R}^3$;
(b) $3 + x^3, 2 - x - x^2, x + x^2 - x^3, x + 2x^2$ for P_3;
(c) $1 + i, 1 - i$ for $\mathbb{C}$ (as a vector space over $\mathbb{R}$);
(d) $\begin{bmatrix} 1 & 0 \\ 0 & 1 \end{bmatrix}, \begin{bmatrix} 0 & 1 \\ 1 & 0 \end{bmatrix}, \begin{bmatrix} 1 & 0 \\ 1 & 0 \end{bmatrix}, \begin{bmatrix} 0 & 1 \\ 1 & -1 \end{bmatrix}$ for $M_{2,2}(\mathbb{R})$.

4 Write $(1, 4, -5)$ as a linear combination of the basis vectors in 3(a) above.

5 Write $1 + 2x + 3x^2 + 4x^3$ as a linear combination of the basis vectors in 3(b) above.

6 Prove Lemma 5.3.2. (If you have difficulties look back at section 1.2.)

7 Prove that any sequence of vectors in a vector space V which includes the zero vector must be linearly dependent.

8 Prove that, in any vector space V, a sequence of vectors in which the same vector occurs more than once is necessarily linearly dependent.

5.4 THE DIMENSION OF A VECTOR SPACE

We saw in section 1.2 that the dimension of $\mathbb{R}^2$, and of $\mathbb{R}^3$, was equal to the number of vectors occurring in a basis for the space. This suggests defining the dimension of a general vector space in the same way. In order for the definition to be sensible, however, we must first ascertain whether every vector space *has* a basis, and whether every two bases for the same vector space contain the same number of elements.

Throughout, V will denote a vector space over a field F. We make life a little easier for ourselves by ensuring that V is not 'too big'. We will call V *finite-dimensional* if there is a finite sequence of vectors in V which span V.

Example 5.4.1

We have seen that the vector spaces $\mathbb{R}^n$, P_n, $\mathbb{C}$, $M_{2,2}(\mathbb{R})$ (and it is not difficult to extend the argument to $M_{m,n}(\mathbb{R})$) are finite-dimensional.

VECTOR SPACES

LEMMA 5.4.1 Let $v \in V$, and suppose that v can be written as a linear combination of $v_1, \ldots, v_n \in V$, and that each v_i can be written as a linear combination of $e_1, \ldots, e_m$. Then v can be written as a linear combination of $e_1, \ldots, e_m$.

Proof Suppose that

$$v = \sum_{i=1}^{n} \alpha_i v_i \quad \text{and that} \quad v_i = \sum_{j=1}^{m} \beta_{ij} e_j.$$

Then

$$v = \sum_{i=1}^{n} \alpha_i v_i = \sum_{i=1}^{n} \alpha_i \left(\sum_{j=1}^{m} \beta_{ij} e_j \right) = \sum_{i=1}^{n} \sum_{j=1}^{m} (\alpha_i \beta_{ij}) e_j.$$

Example 5.4.2

In case you find that the use of summation notation in the above proof makes it difficult to follow, we will illustrate it with a very simple special case. Suppose that

$$v = 2v_1 - 3v_2 \quad \text{and that} \quad v_1 = e_1 - e_2 + 5e_3, \; v_2 = e_1 - e_3.$$

Then

$$v = 2(e_1 - e_2 + 5e_3) - 3(e_1 - e_3) = -e_1 - 2e_2 + 13e_3.$$

THEOREM 5.4.2 Let V be finite-dimensional. Then every finite sequence of vectors which spans V contains a subsequence which is a basis for V.

Proof Let $e_1, \ldots, e_n$ span V. Move along the sequence one at a time, in order from e_1 to e_n, and at each stage discard e_j if it can be written as a linear combination of the e_i's for $i < j$. This produces a subsequence $e_{i_1}, \ldots, e_{i_r}$ (where $i_1 < i_2 < \ldots < i_r$, and $r \leq n$) in which no vector is linearly dependent on its predecessors. Furthermore, by Lemma 5.4.1, any vector which could be written as a linear combination of the original sequence can still be written as a linear combination of this subsequence. The subsequence, therefore, spans V.

It remains to show that it is linearly independent. Suppose to the contrary that it is linearly dependent, so that

$$\alpha_1 e_{i_1} + \ldots + \alpha_r e_{i_r} = 0 \text{ and } \alpha_j \neq 0 \quad \text{for some } 1 \leq j \leq r.$$

Let α_s be the last non-zero element in the sequence $\alpha_1, \ldots, \alpha_r$, so that

$$\alpha_1 e_{i_1} + \ldots + \alpha_s e_{i_s} = 0 \text{ and } \alpha_s \neq 0.$$

We cannot have $s = 1$, as this would imply that $e_{i_1} = 0$, and it could not form

113

part of a basis (exercise 5.3.7). Thus, $s > 1$. But then, rearranging the equation gives $\mathbf{e}_{i_s} = -\alpha_s^{-1}(\alpha_1 \mathbf{e}_{i_1} + \ldots + \alpha_{s-1} \mathbf{e}_{i_{s-1}})$, and $\mathbf{e}_{i_s}$ is linearly dependent on its predecessors. This contradiction completes the proof.

COROLLARY 5.4.3 Let V be finite-dimensional. Then V has a (finite) basis.

Proof By definition, V has a finite spanning sequence, and a subsequence of this forms a basis for V.

Examples 5.4.3

Find a basis for the subspace $W = \langle (1, 1, -2), (2, 1, -3), (-1, 0, 1), (0, 1, -1) \rangle$ of $\mathbb{R}^3$.

Solution Clearly $(0, 1, -1), (-1, 0, 1), (1, 1, -2), (2, 1, -3)$ is a spanning sequence for W. (We could have written down the vectors in any order, and still have had a spanning sequence. In order to simplify the arithmetic we have chosen the order so that the vectors which look 'easiest' to handle appear at the beginning.) We carry out the procedure described in Theorem 5.4.2. We cannot write $(-1, 0, 1)$ as $\alpha(0, 1, -1)$ for any $\alpha \in \mathbb{R}$, so we keep $(0, 1, -1)$ and $(-1, 0, 1)$.

Now suppose that $(1, 1, -2) = \alpha(0, 1, -1) + \beta(-1, 0, 1)$. Then $1 = -\beta, 1 = \alpha$, $-2 = -\alpha + \beta$, and these equations have solutions $\alpha = 1, \beta = -1$, so

$$(1, 1, -2) = (0, 1, -1) - (-1, 0, 1).$$

We discard, therefore, $(1, 1, -2)$. Working similarly we find that

$$(2, 1, -3) = (0, 1, -1) - 2(-1, 0, 1),$$

so we discard $(2, 1, -3)$ also. It follows from Theorem 5.4.2 that $(-1, 0, 1), (0, 1, -1)$ is a basis for W; in particular, W is also equal to $\langle (-1, 0, 1), (0, 1, -1) \rangle$.

THEOREM 5.4.4 Let V be finite-dimensional, and let $\mathbf{v}_1, \ldots, \mathbf{v}_m \in V$ be linearly independent. Then

(a) the sequence can be extended to a basis $\mathbf{v}_1, \ldots, \mathbf{v}_n$ for V $(n \geq m)$;
(b) if $\mathbf{e}_1, \ldots, \mathbf{e}_p$ is a basis for V, then $m \leq p$.

Proof

(a) Since V is finite-dimensional, there is a finite spanning sequence $\mathbf{e}_1, \ldots, \mathbf{e}_k$. Then the sequence $\mathbf{v}_1, \ldots, \mathbf{v}_m, \mathbf{e}_1, \ldots, \mathbf{e}_k$ also spans V.

Now carry out the procedure described in Theorem 5.4.2 of discarding any vector which is linearly dependent on its predecessors. This produces a subsequence which is a basis for V, as in Theorem 5.4.2. Furthermore, none of the $\mathbf{v}_i$s will be discarded as they are linearly independent.

(b) Consider the sequence $\mathbf{v}_m, \mathbf{e}_1, \ldots, \mathbf{e}_p$ which spans V. Carry out the procedure of Theorem 5.4.2 on this sequence. At least one vector must be discarded, as these vectors are linearly dependent ($\mathbf{v}_m$ can be written as a linear combination of $\mathbf{e}_1, \ldots, \mathbf{e}_p$, as can every vector in V), and all of the vectors discarded are $\mathbf{e}_i$s ($1 \leqslant i \leqslant p$). A basis $\mathbf{v}_m, \mathbf{e}_{i_1}, \ldots, \mathbf{e}_{i_r}$ results, with $r \leqslant p-1$.

Now apply the same process to $\mathbf{v}_{m-1}, \mathbf{v}_m, \mathbf{e}_{i_1}, \ldots, \mathbf{e}_{i_r}$. The same reasoning shows that a basis $\mathbf{v}_{m-1}, \mathbf{v}_m, \mathbf{e}_{j_1}, \ldots, \mathbf{e}_{j_s}$ results, with $s \leqslant p-2$.

Continuing in this way we construct eventually a basis $\mathbf{v}_1, \ldots, \mathbf{v}_m, \mathbf{e}_{k_1}, \ldots, \mathbf{e}_{k_t}$, and $t \leqslant p - m$. But, clearly, $t \geqslant 0$, so $p - m \geqslant t \geqslant 0$; that is, $p \geqslant m$.

COROLLARY 5.4.5 Let V be finite-dimensional. Then every basis for V contains the same number of vectors.

Proof Suppose that $\mathbf{e}_1, \ldots, \mathbf{e}_m$ and $\mathbf{f}_1, \ldots, \mathbf{f}_n$ are both bases for V. Then $\mathbf{e}_1, \ldots, \mathbf{e}_m$ are linearly independent, so $m \leqslant n$ by Theorem 5.4.4(b). But $\mathbf{f}_1, \ldots, \mathbf{f}_n$ are also linearly independent, so $n \leqslant m$ by Theorem 5.4.4(b) again.

As a result of Corollaries 5.4.3 and 5.4.5 we can now define the *dimension* of a finite-dimensional vector space V to be the number of vectors in any basis for V; we shall denote it by $\dim V$ (or by $\dim_F V$ if we want to emphasise the field of scalars of V).

Remark Strictly speaking, we have overlooked the case of $V = \{\mathbf{0}\}$ throughout this section. This can be tidied up by defining $\dim\{\mathbf{0}\} = 0$ and adopting the convention that the empty set is a basis for $\{\mathbf{0}\}$. However, $\{\mathbf{0}\}$ is a particularly uninteresting vector space, and so we shall leave the details to be filled in by those interested in doing so.

Examples 5.4.4

1. $\dim \mathbb{R}^n = n$; $\dim P_n = n + 1$; $\dim M_{2,2}(\mathbb{R}) = 4$; $\dim_\mathbb{R} \mathbb{C} = 2$.
2. If $W = \langle (1, 1, -2), (2, 1, -3), (-1, 0, 1), (0, 1, -1) \rangle$ then $\dim W = 2$, as we saw earlier that $(0, 1, -1), (-1, 0, 1)$ is a basis for W.
3. Let W be any subspace of $\mathbb{R}^2$. Then, if $W \neq \{\mathbf{0}\}$, $\dim W = 1$ or 2.
   (a) If $\dim W = 1$, let $\mathbf{e}$ be a basis for W. Then $W = \langle \mathbf{e} \rangle$ which is represented by a line through the origin (example 5.2.2.1).
   (b) If $\dim W = 2$, let $\mathbf{e}, \mathbf{f}$ be a basis for W. Then $W = \langle \mathbf{e}, \mathbf{f} \rangle = \mathbb{R}^2$.

   Thus, the subspaces of $\mathbb{R}^2$ are $\{\mathbf{0}\}$, $\mathbb{R}^2$ and lines through the origin.

4. If W is a subspace of V, then dim $W \leq$ dim V. For, let $\mathbf{w}_1, \ldots, \mathbf{w}_m$ be a basis for W. Then $\mathbf{w}_1, \ldots, \mathbf{w}_m$ is a linearly independent sequence of vectors in V. By Theorem 5.4.4(a) the sequence can be extended to a basis for V. The result is then clear.

EXERCISES 5.4

1. Calculate the dimension of each of the following vector spaces:
   (a) $M_{m,n}(\mathbb{R})$;
   (b) the subspace $\langle (1, 0, -1, 5), (3, 2, 1, 0), (0, -1, 0, 1), (-1, -5, -3, 13) \rangle$ of $\mathbb{R}^4$;
   (c) the subspace $\langle 1+x, 1+x+x^2, 1-x-x^2, 1+2x+x^2 \rangle$ of P_2;
   (d) $\mathbb{C}$ considered as a vector space over $\mathbb{C}$;
   (e) the subspace
   $$\left\langle \begin{bmatrix} 1 & 1 \\ -2 & 1 \end{bmatrix}, \begin{bmatrix} -1 & 0 \\ 3 & -2 \end{bmatrix}, \begin{bmatrix} 0 & 1 \\ 1 & -1 \end{bmatrix}, \begin{bmatrix} 2 & 1 \\ -5 & 3 \end{bmatrix}, \begin{bmatrix} 3 & 2 \\ -7 & 4 \end{bmatrix} \right\rangle$$
   of $M_{2,2}(\mathbb{R})$;
   (f) the subspace $\{(x, y, z, w): x + y - z = y + w = 0\}$ of $\mathbb{R}^4$.

2. Describe geometrically all of the subspaces of $\mathbb{R}^3$.

3. Let V be a vector space of dimension n. Prove that
   (a) any sequence of n vectors which span V is a basis for V;
   (b) any sequence of n vectors which are linearly independent is a basis for V;
   (c) no sequence of less than n vectors can span V;
   (d) every sequence of more than n vectors in V must be linearly dependent.

4. Let W be a subspace of V such that dim $W =$ dim V. Prove that $W = V$.

5. Let U, W be subspaces of the finite-dimensional vector space V. Prove that
   $$\dim(U + W) = \dim U + \dim W - \dim(U \cap W).$$
   (*Hint:* Pick a basis $\mathbf{e}_1, \ldots, \mathbf{e}_r$ for $U \cap W$ and extend it to a basis $\mathbf{e}_1, \ldots, \mathbf{e}_r$, $\mathbf{f}_1, \ldots, \mathbf{f}_s$ for U, and $\mathbf{e}_1, \ldots, \mathbf{e}_r, \mathbf{g}_1, \ldots, \mathbf{g}_t$ for W. Show that $\mathbf{e}_1, \ldots, \mathbf{e}_r, \mathbf{f}_1, \ldots, \mathbf{f}_s$, $\mathbf{g}_1, \ldots, \mathbf{g}_t$ is a basis for $U + W$. See exercise 5.2.8 for definitions of $U \cap W$, $U + W$.)

6. Two subspaces U, W of V are said to be *complementary* if $V = U + W$ and $U \cap W = \{\mathbf{0}\}$. Prove that for every subspace of a finite-dimensional vector space V there is a complementary subspace.

SOLUTIONS AND HINTS FOR EXERCISES

Exercises 5.1

1 Checking the axioms is straightforward. Here $\mathbf{0} = (0,0,0,\ldots)$, $-(a_1, a_2, a_3, \ldots) = (-a_1, -a_2, -a_3, \ldots)$.

2 Here, $\mathbf{0} = 0 + 0\sqrt{2}$, $-(a + b\sqrt{2}) = -a + (-b)\sqrt{2}$.

3 Here $\mathbf{0}$ is the function $f_0: \mathbb{R} \to \mathbb{R}$, where $f_0(x) = 0$ for all $x \in \mathbb{R}$, and $(-f)(x) = -f(x)$ for all $x \in \mathbb{R}$.

4 Note that $0\mathbf{a} = (0 + 0)\mathbf{a} = 0\mathbf{a} + 0\mathbf{a}$ (by V6). Now $0\mathbf{a} \in V$, so $-(0\mathbf{a})$ exists (by V3). Hence

$$\mathbf{0} = 0\mathbf{a} + (-(0\mathbf{a})) \quad \text{(by V3)}$$
$$= (0\mathbf{a} + 0\mathbf{a}) + (-(0\mathbf{a})) = 0\mathbf{a} + (0\mathbf{a} + (-(0\mathbf{a}))) \quad \text{(by V1)}$$
$$= 0\mathbf{a} + \mathbf{0} \quad \text{(by V3)}$$
$$= 0\mathbf{a} \quad \text{(by V2).}$$

5 *Lemma 5.1.1*

(b) Suppose that $\mathbf{u}, \mathbf{w} \in V$ are both such that
$$\mathbf{v} + \mathbf{u} = \mathbf{v} + \mathbf{w} = \mathbf{0} = \mathbf{u} + \mathbf{v} = \mathbf{w} + \mathbf{v}.$$
Then
$$\mathbf{u} = \mathbf{u} + \mathbf{0} \quad \text{(by V2)}$$
$$= \mathbf{u} + (\mathbf{v} + \mathbf{w}) = (\mathbf{u} + \mathbf{v}) + \mathbf{w} \quad \text{(by V1)}$$
$$= \mathbf{0} + \mathbf{w} = \mathbf{w} \quad \text{(by V2).}$$

(e) We have $\mathbf{v} + (-\mathbf{v}) = \mathbf{0} = (-\mathbf{v}) + \mathbf{v}$ by (V3). But $-(-\mathbf{v})$ is that unique element of V which, when added to $-\mathbf{v}$ on either side, yields $\mathbf{0}$. Since $\mathbf{v}$ has this property, $-(-\mathbf{v}) = \mathbf{v}$.

(f) Likewise, $-(\alpha \mathbf{v})$ is the unique element of V which, when added to $\alpha \mathbf{v}$ on either side, gives $\mathbf{0}$. But

$$\alpha(-\mathbf{v}) + (\alpha \mathbf{v}) = \alpha((-\mathbf{v}) + \mathbf{v}) \quad \text{(by V5)}$$
$$= \alpha \mathbf{0} \quad \text{(by V3)}$$
$$= \mathbf{0} \quad \text{by Lemma 5.1.1(c)}$$

Similarly, $(\alpha \mathbf{v}) + \alpha(-\mathbf{v}) = \mathbf{0}$, and the result follows.

(g)
$$(-\alpha)(-\mathbf{v}) = -(\alpha(-\mathbf{v})) \quad \text{by Lemma 5.1.1(d)}$$
$$= -(-(\alpha \mathbf{v})) \quad \text{by Lemma 5.1.1(f)}$$
$$= \alpha \mathbf{v} \quad \text{by Lemma 5.1.1(e).}$$

Exercises 5.2

1. A typical element of $\mathbb{R}^n$ is of the form
$$(x_1, x_2, \ldots, x_n) = x_1\mathbf{v}_1 + x_2\mathbf{v}_2 + \ldots + x_n\mathbf{v}_n, \qquad x_i \in \mathbb{R} \qquad (1 \leq i \leq n).$$

2. $\langle \mathbf{v}_1 \rangle$ represents a straight line through the origin;
   $\langle \mathbf{v}_1, \mathbf{v}_2 \rangle$ represents a plane through the origin;
   $\langle \mathbf{v}_1, \mathbf{v}_2, \mathbf{v}_3 \rangle = \mathbb{R}^3$.

3. In each part denote the given set by S; all of them are clearly non-empty, so S1 is satisfied in each case.
   (a) This is not a subspace. For example, $(1,0,0) \in S$, but
   $$(1,0,0) + (1,0,0) = (2,0,0) \notin S,$$
   so S2 is not satisfied.
   (b) Let $(x, y, z), (x_1, y_1, z_1), (x_2, y_2, z_2) \in S$.
   Then $x - y = z, \; x_1 - y_1 = z_1, \; x_2 - y_2 = z_2$.
   Thus $(x_1, y_1, z_1) + (x_2, y_2, z_2) = (x_1 + x_2, y_1 + y_2, z_1 + z_2)$,
   and $(x_1 + x_2) - (y_1 + y_2) = (z_1 + z_2)$,
   so S2 is satisfied.
   Also, $\alpha(x, y, z) = (\alpha x, \alpha y, \alpha z) \in S$,
   since $\alpha x - \alpha y = \alpha z$, and S3 is satisfied. It follows that S is a subspace.
   (c) Here $S = \{\mathbf{0}\}$, which is a subspace.
   (d) This is not a subspace. For example, $(1,0,0) \in S$, but
   $$(1,0,0) + (1,0,0) = (2,0,0) \notin S,$$
   and S2 is not satisfied.
   (e) This is not a subspace. For example, $(-1, 1, 0) \in S$, but
   $$(-1, 1, 0) + (-1, 1, 0) = (-2, 2, 0) \notin S,$$
   and so S2 is not satisfied.

4. Again denote the given set by S.
   (a) This is not a subspace. For, if $f, g \in S$, then $f(1) = g(1) = 1$, but $(f+g)(1) = f(1) + g(1) = 1 + 1 = 2$, so $f + g \notin S$, and S2 fails.
   (b) This is a subspace. The zero function belongs to S, so S is non-empty and S1 holds. Suppose that $f, g \in S$, so that $f(1) = g(1) = 0$. Then
   $$(f + g)(1) = f(1) + g(1) = 0, \qquad \text{so S2 holds, and}$$
   $$(\alpha f)(1) = \alpha(f(1)) = 0, \qquad \text{so S3 holds.}$$
   (c) This is a subspace. The zero function belongs to S, so S is non-empty and S1 holds. Suppose that $f, g \in S$, so that $f(x) = f(0), g(x) = g(0)$ for all

$x \in \mathbb{R}$. Then

$$(f + g)(x) = f(x) + g(x) = f(0) + g(0) = (f + g)(0), \quad \text{so S2 holds, and}$$
$$(\alpha f)(x) = \alpha f(x) = \alpha f(0) = (\alpha f)(0), \quad \text{so S3 holds.}$$

(d) If $f(x) = f(1)$ for *all* $x \in \mathbb{R}$, then, in particular, $f(0) = f(1)$. It follows that this is the same set as in part (c) above.

(e) This is not a subspace. For the function $f: \mathbb{R} \to \mathbb{R}$ given by $f(x) = x^2$ for all $x \in \mathbb{R}$ belongs to S. But $(-1)f = -f \notin S$, since $-f(1) = -1 < 0$.

5 Let $f(x) = a_0 + a_1 x + \ldots + a_n x^n$. Then $f(0) = a_0$. Hence the given set is $\{a_1 x + \ldots + a_n x^n : a_1, \ldots, a_n \in \mathbb{R}\}$. A spanning sequence is $x, x^2, \ldots, x^n$, for example.

6 A typical element of $M_{2,2}(\mathbb{R})$ is

$$\begin{bmatrix} a & b \\ c & d \end{bmatrix} = a\begin{bmatrix} 1 & 0 \\ 0 & 0 \end{bmatrix} + b\begin{bmatrix} 0 & 1 \\ 0 & 0 \end{bmatrix} + c\begin{bmatrix} 0 & 0 \\ 1 & 0 \end{bmatrix} + d\begin{bmatrix} 0 & 0 \\ 0 & 1 \end{bmatrix}.$$

7 Here

$$\begin{bmatrix} a & b \\ c & d \end{bmatrix} = (a-b)\begin{bmatrix} 1 & 0 \\ 0 & 0 \end{bmatrix} + (b-c)\begin{bmatrix} 1 & 1 \\ 0 & 0 \end{bmatrix} + (c-d)\begin{bmatrix} 1 & 1 \\ 1 & 0 \end{bmatrix} + d\begin{bmatrix} 1 & 1 \\ 1 & 1 \end{bmatrix}.$$

8 (a) $U \cap W$ is not empty, since $\mathbf{0} \in U \cap W$. Let $\mathbf{x}, \mathbf{y} \in U \cap W$. Then $\mathbf{x}, \mathbf{y} \in U$, so $\mathbf{x} + \mathbf{y} \in U$ as U is a subspace; also, $\mathbf{x}, \mathbf{y} \in W$, so $\mathbf{x} + \mathbf{y} \in W$ as W is a subspace. Thus, $\mathbf{x} + \mathbf{y} \in U \cap W$. Moreover, $\alpha \mathbf{x} \in U$ and $\alpha \mathbf{x} \in W$, so $\alpha \mathbf{x} \in U \cap W$.

(b) Again $\mathbf{0} \in U + W$, so $U + W$ is not empty. Let $\mathbf{x}, \mathbf{y} \in U + W$. Then $\mathbf{x} = \mathbf{u} + \mathbf{w}, \mathbf{y} = \mathbf{u}' + \mathbf{w}'$ for some $\mathbf{u}, \mathbf{u}' \in U$, $\mathbf{w}, \mathbf{w}' \in W$.

Hence $\quad \mathbf{x} + \mathbf{y} = (\mathbf{u} + \mathbf{u}') + (\mathbf{w} + \mathbf{w}') \in U + W,$

and $\quad \alpha \mathbf{x} = (\alpha \mathbf{u}) + (\alpha \mathbf{w}) \in U + W.$

(c) If $U \subset W$, then $U \cup W = W$, which is a subspace; if $W \subset U$, then $U \cup W = U$, which is a subspace.

So let $U \cup W$ be a subspace, and suppose that $U \not\subset W$. Then there is an element $\mathbf{u} \in U$ such that $\mathbf{u} \notin W$. Let $\mathbf{w}$ be *any* element of W. Then $\mathbf{u}, \mathbf{w} \in U \cup W$ and so $\mathbf{u} + \mathbf{w} \in U \cup W$, since it is a subspace. If $\mathbf{u} + \mathbf{w} \in W$, then $\mathbf{u} + \mathbf{w} = \mathbf{w}' \in W$, which implies that $\mathbf{u} = \mathbf{w}' - \mathbf{w} \in W$, a contradiction. Hence $\mathbf{u} + \mathbf{w} \in U$; put $\mathbf{u} + \mathbf{w} = \mathbf{u}' \in U$. Then $\mathbf{w} = \mathbf{u}' - \mathbf{u} \in U$. It follows that $W \subset U$.

Exercises 5.3

1 $\begin{bmatrix} 0 & 0 \\ 0 & 0 \end{bmatrix} = a\begin{bmatrix} 1 & 0 \\ 0 & 0 \end{bmatrix} + b\begin{bmatrix} 0 & 1 \\ 0 & 0 \end{bmatrix} + c\begin{bmatrix} 0 & 0 \\ 1 & 0 \end{bmatrix} + d\begin{bmatrix} 0 & 0 \\ 0 & 1 \end{bmatrix} = \begin{bmatrix} a & b \\ c & d \end{bmatrix}$

$\Rightarrow a = b = c = d = 0.$

Hence the matrices are linearly independent; combining this with exercise 5.2.6 shows that they form a basis.

2. (a) $a(1, 0, -1, 1) + b(0, 1, -3, 2) + c(-1, 2, 0, 1) + d(0, 4, 0, -1) = (0, 0, 0, 0)$
$\Rightarrow a - c = 0, b + 2c + 4d = 0, -a - 3b = 0, a + 2b + c - d = 0$
$\Rightarrow a = b = c = d = 0$. Hence the sequence is linearly independent.
(b) $a(1 - x) + b(1 + x) + c(1 - x + 2x^2) = 0 \Rightarrow$
$a + b + c = -a + b - c = 2c = 0 \Rightarrow a = b = c = 0$.
Hence the sequence is linearly independent.
(c) $a(1 + i) + b(1 - i) + c(2 + 3i) = 0 \Rightarrow a + b + 2c = 0 = a - b + 3c$.
Now $a = 5, b = -1, c = -2$ is a non-trivial solution to this system, so the sequence is linearly dependent.
(d) $\begin{bmatrix} 1-1 \\ -1 & 1 \end{bmatrix} + \begin{bmatrix} -1 & 1 \\ 1 & -1 \end{bmatrix} + 0\begin{bmatrix} 1 & 2 \\ 3 & 4 \end{bmatrix} + 0\begin{bmatrix} 0 & -3 \\ -4 & -3 \end{bmatrix} = \begin{bmatrix} 0 & 0 \\ 0 & 0 \end{bmatrix}$,
so the sequence is linearly dependent.

3. (a) $a(1, -2, -1) + b(-1, 0, 3) + c(0, 2, -1) = (0, 0, 0)$
$\Rightarrow a - b = -2a + 2c = -a + 3b - c = 0 \Rightarrow a = b = c = 0$,
so the sequence is linearly independent. Furthermore,
$$(x, y, z) = (3x + \tfrac{1}{2}y + z)(1, -2, -1) + (2x + \tfrac{1}{2}y + z)(-1, 0, 3)$$
$$+ (3x + y + z)(0, 2, -1),$$
so the sequence spans $\mathbb{R}^3$.
(b) $a(3 + x^3) + b(2 - x - x^2) + c(x + x^2 - x^3) + d(x + 2x^2) = 0$
$\Rightarrow 3a + 2b = -b + c + d = -b + c + 2d = a - c = 0$
$\Rightarrow a = b = c = d = 0$, so the sequence is linearly independent. Moreover,
$a + bx + cx^2 + dx^3 = \tfrac{1}{5}(a + 4b - 2c + 2d)(3 + x^3)$
$+ \tfrac{1}{5}(a - 6b + 3c - 3d)(2 - x - x^2) + \tfrac{1}{5}(a + 4b - 2c - 3d)(x + x^2 - x^3)$
$+ (c - b)(x + 2x^2)$, so the sequence spans P_3.
(c) $a(1 + i) + b(1 - i) = 0 \Rightarrow a + b = a - b = 0 \Rightarrow a = b = 0$, so the sequence is linearly independent. Furthermore, $a + ib = \tfrac{1}{2}(a + b)(1 + i) + \tfrac{1}{2}(a - b)(1 - i)$, so the sequence spans $\mathbb{C}$.
(d) $a\begin{bmatrix} 1 & 0 \\ 0 & 1 \end{bmatrix} + b\begin{bmatrix} 0 & 1 \\ 1 & 0 \end{bmatrix} + c\begin{bmatrix} 1 & 0 \\ 1 & 0 \end{bmatrix} + d\begin{bmatrix} 0 & 1 \\ 1 & -1 \end{bmatrix} = \begin{bmatrix} 0 & 0 \\ 0 & 0 \end{bmatrix}$
$\Rightarrow a + c = b + d = b + c + d = a - d = 0 \Rightarrow a = b = c = d = 0$,
so the sequence is linearly independent. Moreover,
$\begin{bmatrix} a & b \\ c & d \end{bmatrix} = (a + b - c)\begin{bmatrix} 1 & 0 \\ 0 & 1 \end{bmatrix} + (c - a + d)\begin{bmatrix} 0 & 1 \\ 1 & 0 \end{bmatrix} + (c - b)\begin{bmatrix} 1 & 0 \\ 1 & 0 \end{bmatrix}$
$+ (a + b - c - d)\begin{bmatrix} 0 & 1 \\ 1 & -1 \end{bmatrix}$, so the sequence spans $M_{2,2}(R)$.

4 $(1, 4, -5) = 0(1, -2, -1) - (-1, 0, 3) + 2(0, 2, -1)$.

5 $1 + 2x + 3x^2 + 4x^3 = \frac{11}{5}(3 + x^3) - \frac{14}{5}(2 - x - x^2)$
$- \frac{9}{5}(x + x^2 - x^3) + (x + 2x^2)$.

6 Every element of V can be written as a linear combination of $v_1, \ldots, v_n$ since they span V. Suppose that

$$v = \alpha_1 v_1 + \ldots + \alpha_n v_n = \beta_1 v_1 + \ldots + \beta_n v_n.$$

Then $\qquad (\alpha_1 - \beta_1)v_1 + \ldots + (\alpha_n - \beta_n)v_n = 0,$

and linear independence implies that $\alpha_1 = \beta_1, \ldots, \alpha_n = \beta_n$.

7 Choose the coefficient of **0** to be non-zero, and all of the other coefficients to be zero. This gives a non-trivial linear combination which is equal to the vector **0**.

8 Suppose that the vector **v** occurs twice. Then take a linear combination in which the coefficient of the first appearance of **v** is 1, of the second appearance of **v** is -1, and all other coefficients are zero. Clearly this non-trivial linear combination is equal to **0**.

Exercises 5.4

1 (a) This has dimension mn; a basis is $M_{11}, \ldots, M_{1n}, \ldots, M_{m1}, \ldots, M_{mn}$ where M_{ij} is the matrix in which the (i, j)-element is 1, and every other element is 0.

(b) $2(1, 0, -1, 5) - (3, 2, 1, 0) + 3(0, -1, 0, 1) = (-1, -5, -3, 13)$,
so the given spanning sequence is linearly dependent. However,

$$a(1, 0, -1, 5) + b(3, 2, 1, 0) + c(0, -1, 0, 1) = (0, 0, 0, 0)$$
$$\Rightarrow \quad a + 3b = 2b - c = -a + b = 5a + c = 0$$
$$\Rightarrow \quad a = b = c = 0,$$

so these three vectors are linearly independent. Hence the given subspace is the same as $\langle (1, 0, -1, 5), (3, 2, 1, 0), (0, -1, 0, 1) \rangle$ which is three-dimensional.

(c) $a + bx + cx^2 = (b - c)(1 + x) + (\frac{1}{2}a - \frac{1}{2}b + c)(1 + x + x^2)$
$+ \frac{1}{2}(a - b)(1 - x - x^2)$, so the given subspace is the whole of P_2, and thus is three-dimensional.

(d) $\mathbb{C}$ is one-dimensional when considered as a vector space over $\mathbb{C}$; a basis is the real number 1. (This spans $\mathbb{C}$ over $\mathbb{C}$, as every complex number can be written as $a1$, where $a \in \mathbb{C}$.)

(e) $\begin{bmatrix} 0 & 1 \\ 1 & -1 \end{bmatrix} = \begin{bmatrix} 1 & 1 \\ -2 & 1 \end{bmatrix} + \begin{bmatrix} -1 & 0 \\ 3 & -2 \end{bmatrix}$,

$\begin{bmatrix} 2 & 1 \\ -5 & 3 \end{bmatrix} = \begin{bmatrix} 1 & 1 \\ -2 & 1 \end{bmatrix} - \begin{bmatrix} -1 & 0 \\ 3 & -2 \end{bmatrix}$,

$\begin{bmatrix} 3 & 2 \\ -7 & 4 \end{bmatrix} = 2\begin{bmatrix} 1 & 1 \\ -2 & 1 \end{bmatrix} - \begin{bmatrix} -1 & 0 \\ 3 & -2 \end{bmatrix}$,

so the given subspace is the same as

$\left\langle \begin{bmatrix} 1 & 1 \\ -2 & 1 \end{bmatrix}, \begin{bmatrix} -1 & 0 \\ 3 & -2 \end{bmatrix} \right\rangle$. It is easy to check that these two vectors are linearly independent, and so the subspace is two-dimensional.

(f) $\{(x, y, z, w): x + y - z = y + w = 0\} = \{(z - y, y, z, -y): y, z \in \mathbb{R}\}$
$= \{y(-1, 1, 0, -1) + z(1, 0, 1, 0): y, z \in \mathbb{R}\}$

and this subspace is spanned by $(-1, 1, 0, -1)$, $(1, 0, 1, 0)$. It is easy to check that these two vectors are linearly independent, so the dimension is again two.

2 Any non-zero subspace of $\mathbb{R}^3$ is spanned by one, two or three vectors. It follows from exercise 5.2.2 that the subspaces of $\mathbb{R}^3$ are $\{\mathbf{0}\}$, lines through the origin, planes through the origin, and $\mathbb{R}^3$ itself.

3 (a) This is immediate from Theorem 5.4.2 and Corollary 5.4.5.
   (b) This is immediate from Theorem 5.4.4(a) and Corollary 5.4.5.
   (c) This is Theorem 5.4.2 again.
   (d) This is Theorem 5.4.4(a) again.

4 We have $W \subset V$ and dim $W =$ dim V. Suppose that $W \neq V$. Then there is an element $\mathbf{v} \in V$ such that $\mathbf{v} \notin W$. Let $\mathbf{w}_1, \ldots, \mathbf{w}_n$ be a basis for W. Then $\mathbf{v}$ is not a linear combination of $\mathbf{w}_1, \ldots, \mathbf{w}_n$, and so the sequence $\mathbf{v}, \mathbf{w}_1, \ldots, \mathbf{w}_n$ is linearly independent. It follows from Theorem 5.4.4(a) that dim $V \geq n + 1$, a contradiction. Hence $W = V$.

5 Let us follow the hint. Pick $\mathbf{x} \in U + W$; then $\mathbf{x} = \mathbf{u} + \mathbf{w}$, where $\mathbf{u} \in U$ and $\mathbf{w} \in W$.
   Now there are $a_1, \ldots, a_r, b_1, \ldots, b_s, c_1, \ldots, c_r, d_1, \ldots, d_t \in F$ (the underlying field)
   such that $\mathbf{u} = a_1 \mathbf{e}_1 + \ldots + a_r \mathbf{e}_r + b_1 \mathbf{f}_1 + \ldots + b_s \mathbf{f}_s$,
   and $\mathbf{w} = c_1 \mathbf{e}_1 + \ldots + c_r \mathbf{e}_r + d_1 \mathbf{g}_1 + \ldots + d_t \mathbf{g}_t$.
   But then
   $$\mathbf{x} = \mathbf{u} + \mathbf{w} = (a_1 + c_1)\mathbf{e}_1 + \ldots + (a_r + c_r)\mathbf{e}_r + b_1 \mathbf{f}_1$$
   $$+ \ldots + b_s \mathbf{f}_s + d_1 \mathbf{g}_1 + \ldots + d_t \mathbf{g}_t,$$

and so $\mathbf{e}_1,\ldots,\mathbf{e}_r,\mathbf{f}_1,\ldots,\mathbf{f}_s,\mathbf{g}_1,\ldots,\mathbf{g}_t$ spans $U + W$.

Suppose that
$$a_1\mathbf{e}_1 + \ldots + a_r\mathbf{e}_r + b_1\mathbf{f}_1 + \ldots + b_s\mathbf{f}_s + c_1\mathbf{g}_1 + \ldots + c_t\mathbf{g}_t = \mathbf{0}.$$
Then
$$a_1\mathbf{e}_1 + \ldots + a_r\mathbf{e}_r + b_1\mathbf{f}_1 + \ldots + b_s\mathbf{f}_s = -c_1\mathbf{g}_1 - \ldots - c_t\mathbf{g}_t \in U \cap W.$$
Hence
$$-c_1\mathbf{g}_1 - \ldots - c_t\mathbf{g}_t = d_1\mathbf{e}_1 + \ldots + d_r\mathbf{e}_r$$
for some $d_1,\ldots,d_r \in F$, and so
$$c_1\mathbf{g}_1 + \ldots + c_t\mathbf{g}_t + d_1\mathbf{e}_1 + \ldots + d_r\mathbf{e}_r = \mathbf{0}.$$
It follows from the linear independence of these vectors that
$$c_1 = \ldots = c_t = d_1 = \ldots = d_r = 0.$$
But now $a_1\mathbf{e}_1 + \ldots + a_r\mathbf{e}_r + b_1\mathbf{f}_1 + \ldots + b_s\mathbf{f}_s = \mathbf{0}$, and it follows from the linear independence of these vectors that $a_1 = \ldots = a_r = b_1 = \ldots = b_s = 0$. We have now established that the sequence $\mathbf{e}_1,\ldots,\mathbf{e}_r,\mathbf{f}_1,\ldots,\mathbf{f}_s,\mathbf{g}_1,\ldots,\mathbf{g}_t$ is linearly independent, and hence that it forms a basis for $U + W$.

6 Let U be a subspace of V. Pick a basis $\mathbf{u}_1,\ldots,\mathbf{u}_r$ for U. Then this is a linearly independent sequence of vectors in V and so can be extended to a basis $\mathbf{u}_1,\ldots,\mathbf{u}_r,\mathbf{w}_1,\ldots,\mathbf{w}_s$ for V, by Theorem 5.4.4. Let W be the subspace spanned by $\mathbf{w}_1,\ldots,\mathbf{w}_s$. Then W and U are complementary subspaces of V.

6 LINEAR TRANSFORMATIONS

6.1 INTRODUCTION

In the study of any abstract mathematical system, of fundamental importance are the mappings which preserve the structure which is the focus of attention in the system. In the case of vector spaces this structure is the addition and the scalar multiplication, and the resulting mappings turn out to be intimately related to matrices. Consequently, a deeper understanding of matrices and important matrix questions results from their examination.

Let V, W be vector spaces over the same field F. A mapping $T: V \to W$ is called a *linear transformation* (also sometimes called a *linear mapping, linear map, homomorphism,* or *vector space morphism*) if it satisfies the following two conditions:

T1 $T(\mathbf{v} + \mathbf{w}) = T(\mathbf{v}) + T(\mathbf{w})$ for all $\mathbf{v}, \mathbf{w} \in V$.
T2 $T(\alpha \mathbf{v}) = \alpha T(\mathbf{v})$ for all $\alpha \in F$,   and all $\mathbf{v} \in V$.

Examples 6.1

1. Let $T: \mathbb{R}^2 \to \mathbb{R}^2$ be defined by $T((x, y)) = (x - y, 2y)$. Then
$$T((x_1, y_1) + (x_2, y_2)) = T((x_1 + x_2, y_1 + y_2))$$
$$= (x_1 + x_2 - (y_1 + y_2), 2(y_1 + y_2)) = (x_1 - y_1, 2y_1) + (x_2 - y_2, 2y_2)$$
$$= T((x_1, y_1)) + T((x_2, y_2)), \text{ so T1 is satisfied.}$$

Also,
$$T(\alpha(x, y)) = T((\alpha x, \alpha y)) = (\alpha x - \alpha y, 2\alpha y) = \alpha(x - y, 2y) = \alpha T((x, y)),$$
so T2 is satisfied, and T is a linear transformation.

2. Let $D: P_n \to P_{n-1}$ be defined by

$$D(a_0 + a_1x + \ldots + a_nx^n) = a_1 + 2a_2x + \ldots + na_nx^{n-1}.$$

Then
$$D((a_0 + a_1x + \ldots + a_nx^n) + (b_0 + b_1x + \ldots + b_nx^n))$$
$$= D((a_0 + b_0) + (a_1 + b_1)x + \ldots + (a_n + b_n)x^n)$$
$$= (a_1 + b_1) + \ldots + n(a_n + b_n)x^{n-1}$$
$$= (a_1 + \ldots + na_nx^{n-1}) + (b_1 + \ldots + nb_nx^{n-1})$$
$$= D(a_0 + a_1x + \ldots + a_nx^n) + D(b_0 + b_1x + \ldots + b_nx^n),$$

so T1 is satisfied. Also,
$$D(\alpha(a_0 + a_1x + \ldots + a_nx^n)) = D(\alpha a_0 + \alpha a_1x + \ldots + \alpha a_nx^n)$$
$$= \alpha a_1 + \ldots + n\alpha a_nx^{n-1} = \alpha(a_1 + \ldots + na_nx^{n-1})$$
$$= \alpha D(a_0 + a_1x + \ldots + a_nx^n), \text{ and T2 is satisfied.}$$

Hence D is a linear transformation.

3. Let $T: \mathbb{R}^2 \to \mathbb{R}^2$ be defined by $T((x, y)) = (x^2, y)$.
Then
$$T((x_1, y_1) + (x_2, y_2)) = T((x_1 + x_2, y_1 + y_2)) = ((x_1 + x_2)^2, y_1 + y_2)$$
and $T((x_1, y_1)) + T((x_2, y_2)) = (x_1^2, y_1) + (x_2^2, y_2) = (x_1^2 + x_2^2, y_1 + y_2).$

But $(x_1 + x_2)^2 \neq x_1^2 + x_2^2$ in general, so T1 is not satisfied. To put the matter beyond all doubt, note that $(1 + 1)^2 = 4 \neq 2 = 1^2 + 1^2$; so, for example,
$$T((1, 0) + (1, 0)) = T((2, 0)) = (4, 0),$$
whereas $\quad T((1, 0)) + T((1, 0)) = (1, 0) + (1, 0) = (2, 0).$

4. Let $T: \mathbb{R}^2 \to \mathbb{R}^2$ be defined by $T((x, y)) = (x + 1, y)$.
Then $\quad T((x_1 + x_2, y_1 + y_2)) = (x_1 + x_2 + 1, y_1 + y_2)$
whereas
$$T((x_1, y_1)) + T((x_2, y_2)) = (x_1 + 1, y_1) + (x_2 + 1, y_2) = (x_1 + x_2 + 2, y_1 + y_2).$$

Since $x_1 + x_2 + 1 \neq x_1 + x_2 + 2$ for any values of x_1, x_2, T1 is not satisfied, and this transformation is again not linear.

5. Let V be any vector space. The maps $O, J: V \to V$ defined by
$$O(\mathbf{v}) = \mathbf{0}, \quad J(\mathbf{v}) = \mathbf{v}$$
for all $\mathbf{v} \in V$, are easily seen to be linear transformations.

Two easy consequences of the definition of linear transformation are given in the following lemma.

LEMMA 6.1.1 Let V, W be vector spaces over the same field F, and let $T: V \to W$ be a linear transformation. Then (a) $T(\mathbf{0}) = \mathbf{0}$, and (b) $T(-\mathbf{v}) = -T(\mathbf{v})$ for all $\mathbf{v} \in V$.

Proof Simply put (a) $\alpha = 0$, (b) $\alpha = -1$ in T2.

Let V, W, X be vector spaces over the same field F. We will denote by $L(V, W)$ the set of linear transformations from V to W. Then $L(V, W)$ can itself be given the structure of a vector space. If $T, T_1, T_2 \in L(V, W)$, and $\alpha \in F$, then we define $T_1 + T_2$, $\alpha T \in L(V, W)$ by

$$(T_1 + T_2)(\mathbf{v}) = T_1(\mathbf{v}) + T_2(\mathbf{v}), \quad (\alpha T)(\mathbf{v}) = \alpha(T(\mathbf{v})) \quad \text{for all } \mathbf{v} \in V.$$

(Check that $T_1 + T_2$ and αT are linear transformations.) Putting $-T = (-1)T$, and letting O be defined as in example 6.1.5 above, it is easy to check that axioms V1 to V9 are satisfied.

If $T_1 \in L(V, W)$ and $T_2 \in L(W, X)$ then the *composite*, or *product*, can be defined as usual by $T_2 T_1(\mathbf{v}) = T_2(T_1(\mathbf{v}))$ for all $\mathbf{v} \in V$ (Fig. 6.1).

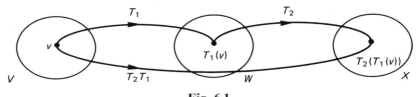

Fig. 6.1

Suppose that $\mathbf{v}_1, \mathbf{v}_2, \mathbf{v} \in V$, $\alpha \in F$. Then

$$\begin{aligned}
T_2 T_1(\mathbf{v}_1 + \mathbf{v}_2) &= T_2(T_1(\mathbf{v}_1 + \mathbf{v}_2)) \\
&= T_2(T_1(\mathbf{v}_1) + T_1(\mathbf{v}_2)) && \text{since } T_1 \text{ is linear} \\
&= T_2(T_1(\mathbf{v}_1)) + T_2(T_1(\mathbf{v}_2)) && \text{since } T_2 \text{ is linear} \\
&= T_2 T_1(\mathbf{v}_1) + T_2 T_1(\mathbf{v}_2),
\end{aligned}$$

and
$$\begin{aligned}
T_2 T_1(\alpha \mathbf{v}) &= T_2(T_1(\alpha \mathbf{v})) \\
&= T_2(\alpha T_1(\mathbf{v})) && \text{since } T_1 \text{ is linear} \\
&= \alpha(T_2(T_1(\mathbf{v}))) && \text{since } T_2 \text{ is linear} \\
&= \alpha T_2 T_1(\mathbf{v}).
\end{aligned}$$

Thus, $T_2 T_1 \in L(V, X)$. It is straightforward to check that the properties corresponding to M1 to M6 of section 2.4 hold here.

EXERCISES 6.1

1. Which of the following mappings $T: \mathbb{R}^2 \to \mathbb{R}^2$ are linear transformations?
   (a) $T((x, y)) = (y, 3x - 2y + 1)$.
   (b) $T((x, y)) = (0, x + y)$.
   (c) $T((x, y)) = (x^2, x + y)$.
   (d) $T((x, y)) = (y, x + 3) - 3(x, 1)$.

2. Write down the image of $(2, -1)$ under each of the mappings in 1 above.

3. Show that the mapping $T: \mathbb{C} \to \mathbb{C}$ given by $T(a + bi) = a - bi$ is a linear transformation.

4. Show that the mapping $T: \mathbb{C} \to M_{2,2}(\mathbb{R})$ given by
$$T(a + bi) = \begin{bmatrix} a & a+b \\ a-b & b \end{bmatrix}$$
is a linear transformation.

5. Let $\mathscr{S}$ be the vector space of all infinite sequences of real numbers (see exercise 5.1.1). Show that the mapping $T: \mathscr{S} \to \mathscr{S}$ given by $T((a_1, a_2, a_3, \ldots)) = (0, a_1, a_2, \ldots)$ is a linear transformation.

6. Let V, W be vector spaces over a field F, let $T, T_1, T_2 \in L(V, W)$, and let $\lambda \in F$. Prove that $T_1 + T_2 \in L(V, W)$ and that $\lambda T \in L(V, W)$.

7. Show that a linear transformation $T: V \to W$ is completely determined by its effect on a basis for V.

8. Prove that a mapping $T: V \to W$ is a linear transformation if and only if $T(\alpha \mathbf{v} + \beta \mathbf{w}) = \alpha T(\mathbf{v}) + \beta T(\mathbf{w})$ for all $\alpha, \beta \in F$, and all $\mathbf{v}, \mathbf{w} \in V$.

9. Let U be a subspace of V, and let $T: V \to W$ be a linear transformation. Prove that $T(U) = \{T(\mathbf{u}): \mathbf{u} \in U\}$ is a subspace of W.

6.2 INVERTIBLE LINEAR TRANSFORMATIONS

Let V, W be vector spaces over the same field F. If $T \in L(V, W)$ is bijective, then $T^{-1}: W \to V$ exists and is defined by
$$T^{-1}(\mathbf{w}) = \mathbf{v} \Leftrightarrow T(\mathbf{v}) = \mathbf{w}.$$
Suppose that $\mathbf{w}, \mathbf{w}_1, \mathbf{w}_2 \in W$, and that $\alpha \in F$.
Put $T^{-1}(\mathbf{w}) = \mathbf{v}$, $T^{-1}(\mathbf{w}_1) = \mathbf{v}_1$, $T^{-1}(\mathbf{w}_2) = \mathbf{v}_2$; then $\mathbf{w} = T(\mathbf{v})$, $\mathbf{w}_1 = T(\mathbf{v}_1)$, $\mathbf{w}_2 = T(\mathbf{v}_2)$.

Now
$$T(\mathbf{v}_1 + \mathbf{v}_2) = T(\mathbf{v}_1) + T(\mathbf{v}_2) = \mathbf{w}_1 + \mathbf{w}_2,$$
so
$$T^{-1}(\mathbf{w}_1) + T^{-1}(\mathbf{w}_2) = \mathbf{v}_1 + \mathbf{v}_2 = T^{-1}(\mathbf{w}_1 + \mathbf{w}_2),$$

and T1 holds. Also, $T(\alpha\mathbf{v}) = \alpha T(\mathbf{v}) = \alpha\mathbf{w}$, so $\alpha T^{-1}(\mathbf{w}) = \alpha\mathbf{v} = T^{-1}(\alpha\mathbf{w})$, and T2 is satisfied. We have proved the following:

LEMMA 6.2.1 Let $T \in L(V, W)$ be bijective. Then $T^{-1} \in L(W, V)$.

It is straightforward to check that the properties corresponding to axioms I1, I2 and I3 of section 2.6 hold.

Example 6.2

Let $T: \mathbb{R}^2 \to \mathbb{R}^2$ be defined by $T((x, y)) = (x + y, x - y)$. Prove that T is bijective, and find T^{-1}.

Solution The inverse of a mapping exists if and only if it is bijective, so it suffices to show that T^{-1} exists and to find it. Now T^{-1} is defined by

$$T((x, y)) = (a, b) \Leftrightarrow T^{-1}((a, b)) = (x, y).$$

But $T((x, y)) = (a, b) \Leftrightarrow a = x + y, b = x - y$. So, what is required is for us to write x, y in terms of a, b. We have

$$a + b = (x + y) + (x - y) = 2x, \qquad a - b = (x + y) - (x - y) = 2y,$$

so $x = \tfrac{1}{2}(a + b), y = \tfrac{1}{2}(a - b)$. Thus, T^{-1} is given by

$$T^{-1}((a, b)) = (\tfrac{1}{2}(a + b), \tfrac{1}{2}(a - b)) \qquad \text{for all } a, b \in \mathbb{R}.$$

Invertible linear transformations faithfully carry the structure of one vector space onto another. If there is an invertible linear transformation from a vector space V onto a vector space W (over the same field), then, as vector spaces, V and W are identical: we write $V \simeq W$ in this situation. In particular, V and W will have the same dimension, and a basis for V will be mapped to a basis for W.

THEOREM 6.2.2 Let V, W be finite-dimensional vector spaces over the same field F, let $\mathbf{e}_1, \ldots, \mathbf{e}_n$ be a basis for V and let $T \in L(V, W)$ be invertible. Then $T(\mathbf{e}_1), \ldots, T(\mathbf{e}_n)$ is a basis for W; in particular, $\dim V = \dim W$.

Proof We need to check linear independence and spanning.

Linear independence Let $\alpha_1, \ldots, \alpha_n \in F$. Then

$$\alpha_1 T(\mathbf{e}_1) + \ldots + \alpha_n T(\mathbf{e}_n) = \mathbf{0} \Rightarrow T(\alpha_1 \mathbf{e}_1 + \ldots + \alpha_n \mathbf{e}_n) = \mathbf{0}$$

$$\Rightarrow \alpha_1 \mathbf{e}_1 + \ldots + \alpha_n \mathbf{e}_n = \mathbf{0} \qquad \text{since } T \text{ is injective}$$

$$\Rightarrow \alpha_1 = \ldots = \alpha_n = 0$$

since $\mathbf{e}_1, \ldots, \mathbf{e}_n$ is a basis.

Spanning Let $\mathbf{w} \in W$. Then there is a vector $\mathbf{v} \in V$ such that $T(\mathbf{v}) = \mathbf{w}$, since T is surjective. Now $\mathbf{v} = \alpha_1 \mathbf{e}_1 + \ldots + \alpha_n \mathbf{e}_n$ for some $\alpha_1, \ldots, \alpha_n \in F$, because $\mathbf{e}_1, \ldots, \mathbf{e}_n$ is a basis for V. But then

$$\mathbf{w} = T(\mathbf{v}) = T(\alpha_1 \mathbf{e}_1 + \ldots + \alpha_n \mathbf{e}_n) = \alpha_1 T(\mathbf{e}_1) + \ldots + \alpha_n T(\mathbf{e}_n).$$

The next result is a little startling at first sight. It says, essentially, that for each field F and each positive integer n, there is exactly one vector space of dimension n over F, namely $F^n = \{(x_1, \ldots, x_n): x_i \in F\}$ (cf. $\mathbb{R}^n$). All of the other finite-dimensional vector spaces, such as P_n, are actually disguised versions of F^n; *as* vector spaces, they are the *same* as F^n.

THEOREM 6.2.3 Let V, W be vector spaces of the same dimension over the same field F, let $\mathbf{e}_1, \ldots, \mathbf{e}_n$ be a basis for V, and let $\mathbf{f}_1, \ldots, \mathbf{f}_n \in W$. Then the mapping

$$T\left(\sum_{i=1}^n \alpha_i \mathbf{e}_i\right) = \sum_{i=1}^n \alpha_i \mathbf{f}_i$$

is a linear transformation. If, furthermore, $\mathbf{f}_1, \ldots, \mathbf{f}_n$ is a basis for W, then T is invertible.

Proof Let

$$\mathbf{v}_1 = \sum_{i=1}^n \lambda_i \mathbf{e}_i, \qquad \mathbf{v}_2 = \sum_{i=1}^n \mu_i \mathbf{e}_i, \qquad \alpha \in F.$$

Then

$$T(\mathbf{v}_1 + \mathbf{v}_2) = T\left(\sum_{i=1}^n (\lambda_i + \mu_i)\mathbf{e}_i\right) = \sum_{i=1}^n (\lambda_i + \mu_i)\mathbf{f}_i$$

$$= \sum_{i=1}^n \lambda_i \mathbf{f}_i + \sum_{i=1}^n \mu_i \mathbf{f}_i = T(\mathbf{v}_1) + T(\mathbf{v}_2),$$

and

$$T(\alpha \mathbf{v}_1) = T\left(\sum_{i=1}^n (\alpha \lambda_i)\mathbf{e}_i\right) = \sum_{i=1}^n (\alpha \lambda_i)\mathbf{f}_i = \alpha \sum_{i=1}^n \lambda_i \mathbf{f}_i = \alpha T(\mathbf{v}_1).$$

Hence T is a linear transformation.

Now let $\mathbf{f}_1, \ldots, \mathbf{f}_n$ be a basis for W, and let $\mathbf{w} \in W$.

Then, for some $\alpha_i \in F$, $\mathbf{w} = \sum_{i=1}^n \alpha_i \mathbf{f}_i$, and if $\mathbf{v} = \sum_{i=1}^n \alpha_i \mathbf{e}_i$, then $T(\mathbf{v}) = \mathbf{w}$. Hence T is surjective.

Also, $T(\mathbf{v}_1) = T(\mathbf{v}_2) \Rightarrow \sum_{i=1}^{n} \lambda_i \mathbf{f}_i = \sum_{i=1}^{n} \mu_i \mathbf{f}_i$

$\Rightarrow \sum_{i=1}^{n} (\lambda_i - \mu_i) \mathbf{f}_i = \mathbf{0}$

$\Rightarrow \lambda_i - \mu_i = 0 \quad$ for $1 \leq i \leq n$,

because the $\mathbf{f}_i$s are linearly independent

$\Rightarrow \lambda_i = \mu_i \quad$ for $1 \leq i \leq n$

$\Rightarrow \mathbf{v}_1 = \mathbf{v}_2$.

It follows that T is injective, and thus bijective.

COROLLARY 6.2.4 Let V be a vector space of dimension n over a field F. Then $V \simeq F^n$.

Proof We have simply to note that V and F^n have the same dimension, and use Theorem 6.2.3.

EXERCISES 6.2

1 Prove that the following linear transformations T are bijective, and in each case find T^{-1}:
   (a) $T: \mathbb{R}^3 \to \mathbb{R}^3$ given by $T((x, y, z)) = (y - z, x + y, x + y + z)$.
   (b) $T: \mathbb{C} \to \mathbb{C}$ given by $T(x + yi) = x - yi$.
   (c) $T: P_2 \to P_2$ given by $T(a_0 + a_1 x + a_2 x^2) = a_1 + a_2 x + a_0 x^2$.

2 Given that $T: \mathbb{R}^4 \to M_{2,2}(\mathbb{R})$ given by $T((x, y, z, w)) = \begin{bmatrix} x & y \\ z & w \end{bmatrix}$ is a linear transformation and that $(1, 0, 0, 0)$, $(1, 1, 0, 0)$, $(1, 1, 1, 0)$, $(1, 1, 1, 1)$ is a basis for $\mathbb{R}^4$, use Theorem 6.2.2 to write down a basis for $M_{2,2}(\mathbb{R})$.

3 Use Theorem 6.2.3 to write down a linear transformation $T: \mathbb{R}^3 \to \mathbb{R}^3$ such that $T((1, 0, 0)) = (1, 0, 0)$, $T((0, 1, 0)) = (1, 1, 0)$, $T((0, 0, 1)) = (1, 1, 1)$. Is T invertible?

4 Write down an invertible linear transformation from P_3 to $M_{2,2}$.

6.3 THE MATRIX OF A LINEAR TRANSFORMATION

Let V, W be vector spaces of dimension m and n respectively over the same field

F, and let $\mathbf{e}_1,\ldots,\mathbf{e}_m$ be a basis for V. If T is a linear transformation from V to W, then T is completely determined by its effect on this basis. For, if $\mathbf{v}\in V$, then

$$\mathbf{v} = \sum_{i=1}^{m} \alpha_i \mathbf{e}_i \quad \text{for some } \alpha_i \in F,$$

and
$$T(\mathbf{v}) = T\left(\sum_{i=1}^{m} \alpha_i \mathbf{e}_i\right) = \sum_{i=1}^{m} \alpha_i T(\mathbf{e}_i).$$

Now let $\mathbf{f}_1,\ldots,\mathbf{f}_n$ be a basis for W. Then $T(\mathbf{e}_i) \in W$ for $1 \leq i \leq m$, so

$$\left. \begin{aligned} T(\mathbf{e}_1) &= \alpha_{11}\mathbf{f}_1 + \ldots + \alpha_{n1}\mathbf{f}_n \\ T(\mathbf{e}_2) &= \alpha_{12}\mathbf{f}_1 + \ldots + \alpha_{n2}\mathbf{f}_n \\ &\cdots\cdots\cdots\cdots\cdots\cdots\cdots \\ T(\mathbf{e}_m) &= \alpha_{1m}\mathbf{f}_1 + \ldots + \alpha_{nm}\mathbf{f}_n \end{aligned} \right\} \quad (1)$$

where the α_{ij}s belong to F. We associate with T the matrix of coefficients

$$A = \begin{bmatrix} \alpha_{11} & \alpha_{12} & \cdots & \alpha_{1m} \\ \cdots & \cdots & \cdots & \cdots \\ \alpha_{n1} & \alpha_{n2} & \cdots & \alpha_{nm} \end{bmatrix}.$$

Note: The matrix A has n rows ($n = \dim W$) and m columns ($m = \dim V$). This matrix is uniquely determined by T and by the bases chosen for V, W.

Examples 6.3.1

1. Let $T: \mathbb{R}^3 \to \mathbb{R}^2$ be defined by $T((x,y,z)) = (2x+y, x-y)$, and choose basis $(1,0,0), (0,1,0), (0,0,1)$ for $\mathbb{R}^3$ and $(1,0), (0,1)$ for $\mathbb{R}^2$. Then

$$T((1,0,0)) = (2,1) = 2(1,0) + 1(0,1),$$
$$T((0,1,0)) = (1,-1) = 1(1,0) - 1(0,1),$$
$$T((0,0,1)) = (0,0) = 0(1,0) + 0(0,1),$$

    so the matrix associated with T is $A = \begin{bmatrix} 2 & 1 & 0 \\ 1 & -1 & 0 \end{bmatrix}$.

    Note: A is $2(=\dim \mathbb{R}^2) \times 3(=\dim \mathbb{R}^3)$.

2. The matrix associated with a linear transformation depends on the bases chosen for the domain and codomain. Remember also that a basis is an *ordered* sequence: writing the vectors in a different order gives a different basis. The importance of this remark can be seen by considering the transformation defined in example 1 above again, but changing the

codomain basis to $(0, 1), (1, 0)$. Check that the matrix associated with T then becomes

$$B = \begin{bmatrix} 1 & -1 & 0 \\ 2 & 1 & 0 \end{bmatrix} \neq A.$$

3. Let $T: P_3 \to \mathbb{C}$ be defined by $T(f(x)) = f(i)$, and choose basis $1, x, x^2, x^3$ for P_3, and $1, i$ for $\mathbb{C}$. Then

$$\begin{aligned} T(1) &= 1 & &= 1(1) + 0(i), \\ T(x) &= i & &= 0(1) + 1(i), \\ T(x^2) &= i^2 = -1 & &= -1(1) + 0(i), \\ T(x^3) &= i^3 = -i & &= 0(1) - 1(i), \end{aligned}$$

so the matrix associated with T is

$$C = \begin{bmatrix} 1 & 0 & -1 & 0 \\ 0 & 1 & 0 & -1 \end{bmatrix}.$$

Now let $v \in V$ and suppose that $v = \sum_{i=1}^m \lambda_i e_i$. Then $(\lambda_1, \ldots, \lambda_m)$ is uniquely determined by v. We call $\lambda_1, \ldots, \lambda_m$ the *coordinates* of v with respect to the basis $e_1, \ldots, e_m$. Let us examine what happens to the coordinates of v when it is mapped by the transformation T. As in the previous section,

$$T(v) = T\left(\sum_{i=1}^m \lambda_i e_i\right) = \sum_{i=1}^m \lambda_i T(e_i) = \sum_{i=1}^m \lambda_i \sum_{j=1}^n \alpha_{ji} f_j = \sum_{i=1}^m \sum_{j=1}^n \lambda_i \alpha_{ji} f_j.$$

The coordinates of $T(v)$ with respect to the basis $f_1, \ldots, f_n$ are, therefore,

$$\left(\sum_{i=1}^m \alpha_{1i}\lambda_i, \ldots, \sum_{i=1}^m \alpha_{ni}\lambda_i\right) = (a_1 \cdot \lambda, \ldots, a_n \cdot \lambda)$$

where $\lambda = (\lambda_1, \ldots, \lambda_m)$, $a_j = (\alpha_{j1}, \ldots, \alpha_{jm})$ for $1 \leq j \leq n$.

Hence the coordinates of $T(v)$ can be found from the coordinates of v by performing a matrix multiplication, as follows: the coordinates of $T(v)$ are

$$\begin{bmatrix} a_1 \cdot \lambda \\ \vdots \\ a_n \cdot \lambda \end{bmatrix} = \begin{bmatrix} \alpha_{11} & \alpha_{12} & \cdots & \alpha_{1m} \\ \cdots\cdots\cdots\cdots\cdots\cdots \\ \alpha_{n1} & \alpha_{n2} & \cdots & \alpha_{nm} \end{bmatrix} \begin{bmatrix} \lambda_1 \\ \vdots \\ \lambda_m \end{bmatrix} = A\lambda^T.$$

(Here we have written the coordinates as column vectors. It does not really matter whether we use rows or columns: they carry the same information. We could have used rows by reversing the order of the matrix multiplication, as

$$[a_1 \cdot \lambda \quad \cdots \quad a_n \cdot \lambda] = (A\lambda^T)^T = \lambda A^T.)$$

Conversely, given a matrix A and bases for V, W, a unique linear transformation from V to W is defined by the equations (1).

Examples 6.3.2

1. Let $A = \begin{bmatrix} 1 & -1 \\ 0 & 1 \\ 2 & 3 \end{bmatrix}$, and choose basis $(1,0), (0,1)$ for $\mathbb{R}^2$, and basis $1, x, x^2$ for P_2. Then the associated linear transformation T acts on the basis for $\mathbb{R}^2$ as follows:
$$T((1,0)) = 1(1) + 0(x) + 2(x^2) = 1 + 2x^2,$$
$$T((0,1)) = -1(1) + 1(x) + 3(x^2) = -1 + x + 3x^2.$$
Thus,
$$T((a,b)) = T(a(1,0) + b(0,1)) = aT((1,0)) + bT((0,1))$$
$$= a(1 + 2x^2) + b(-1 + x + 3x^2)$$
$$= (a - b) + bx + (2a + 3b)x^2.$$

2. We could also perform the above calculation by using the coordinates. With respect to the given basis for $\mathbb{R}^2$, the coordinates of (a, b) are the elements of (a, b) (since $(a, b) = a(1, 0) + b(0, 1)$). Hence, the coordinates of $T(\mathbf{v})$ are given by
$$\begin{bmatrix} 1 & -1 \\ 0 & 1 \\ 2 & 3 \end{bmatrix} \begin{bmatrix} a \\ b \end{bmatrix} = \begin{bmatrix} a - b \\ b \\ 2a + 3b \end{bmatrix};$$
that is, they are the elements of $(a - b, b, 2a + 3b)$.
Thus, $$T((a,b)) = (a - b)1 + bx + (2a + 3b)x^2.$$

THEOREM 6.3.1 Let U, V, W be vector spaces over the same field F, let $T_1, T_2, T: U \to V, S: V \to W$ be linear transformations, and suppose that they are associated with matrices A_1, A_2, A, B respectively with respect to certain bases for U, V, W. Then
 (a) $T_1 + T_2$ is associated with $A_1 + A_2$;
 (b) λT is associated with λA ($\lambda \in F$);
 (c) $S_o T$ is associated with BA;
 (d) T is invertible if and only if A is invertible, and in this case T^{-1} is associated with A^{-1}.

(*Note:* All of these are with respect to the *same* 'certain bases' referred to in the first sentence of the statement of this result.)

Proof

(a) and (b) are easy and will be set as exercise 6.3.4.
(c) Let the given bases be $\mathbf{e}_1,\ldots,\mathbf{e}_r$ for U, $\mathbf{f}_1,\ldots,\mathbf{f}_s$ for V, and $\mathbf{g}_1,\ldots,\mathbf{g}_t$ for W, and suppose that $A = [a_{ij}]_{s \times r}$, $B = [b_{ij}]_{t \times s}$. Then

$$T(\mathbf{e}_j) = \sum_{i=1}^{s} a_{ij}\mathbf{f}_i, \qquad S(\mathbf{f}_i) = \sum_{k=1}^{t} b_{ki}\mathbf{g}_k,$$

and so $S_\circ T(\mathbf{e}_j) = S(T(\mathbf{e}_j)) = S\left(\sum_{i=1}^{s} a_{ij}\mathbf{f}_i\right) = \sum_{i=1}^{s} a_{ij}S(\mathbf{f}_i)$

$$= \sum_{i=1}^{s}\sum_{k=1}^{t} a_{ij}b_{ki}\mathbf{g}_k = \sum_{k=1}^{t}\left(\sum_{i=1}^{s} b_{ki}a_{ij}\right)\mathbf{g}_k$$

$$= \sum_{k=1}^{t} c_{kj}\mathbf{g}_k, \qquad \text{where } BA = [c_{kj}]_{t \times r}$$

(see section 2.3).

It follows that $S_\circ T$ is associated with BA.

(d) Suppose first that T is invertible, so that $T^{-1}: V \to U$ exists such that $T_\circ T^{-1} = J_V$, where $J_V: V \to V$ is the identity map on V. Now, $J_V(\mathbf{f}_i) = \mathbf{f}_i$ $(1 \leqslant i \leqslant s)$, so J_V is associated with the identity matrix I. Thus, applying part (c) we have $AC = I$, where C is the matrix associated with T^{-1}. Similarly, $T_\circ^{-1} T = J_U$ (the identity map on U) gives $CA = I$. It follows that A is invertible, and that $C = A^{-1}$.

Now suppose that A is invertible with inverse A^{-1}. With respect to the bases $\mathbf{f}_1,\ldots,\mathbf{f}_s$ and $\mathbf{e}_1,\ldots,\mathbf{e}_r$, A^{-1} is associated with the linear transformation $R: V \to U$. By (c) above, $R_\circ T$ is associated with $A^{-1}A = I$, and so $R_\circ T(\mathbf{v}) = \mathbf{v}$ for all $\mathbf{v} \in U$. Hence $R_\circ T = J_U$. Similarly, $T_\circ R = J_V$, from which it is apparent that T is invertible and $R = T^{-1}$.

EXERCISES 6.3

1 For each of the following linear transformations find the matrix associated with them with respect to the given bases:

(a) $T: \mathbb{R}^2 \to \mathbb{R}^3$ given by $T((x, y)) = (2x - y, x + 3y, -x)$; basis $(1, 0), (0, 1)$ for $\mathbb{R}^2$, and basis $(0, 0, 1), (0, 1, 0), (1, 0, 0)$ for $\mathbb{R}^3$;

(b) $T: \mathbb{R}^4 \to P_1$ given by $T((a, b, c, d)) = (a + b) + (c + d)x$; standard basis for $\mathbb{R}^4$, and basis $1, x$ for P_1;

(c) $T: \mathbb{R}^3 \to \mathbb{R}^3$ given by $T((x, y, z)) = (x, x + y, x + y + z)$; basis $(1, 0, 0)$, $(1, 1, 0), (1, 1, 1)$ in domain, and basis $(1, -1, 0), (-1, -1, -1), (0, 0, 1)$ in codomain;

(d) $T: M_{2,2} \to \mathbb{R}^2$ given by $T\left(\begin{bmatrix} a & b \\ c & d \end{bmatrix}\right) = (a + d, b - c)$; basis $\begin{bmatrix} 1 & 0 \\ 0 & 0 \end{bmatrix}$,

$\begin{bmatrix} 0 & 1 \\ 0 & 0 \end{bmatrix}, \begin{bmatrix} 0 & 0 \\ 1 & 0 \end{bmatrix}, \begin{bmatrix} 0 & 0 \\ 0 & 1 \end{bmatrix}$ for $M_{2,2}$, and standard basis for $\mathbb{R}^2$.

2 Choose the options which describe the following matrices:

(a) $\begin{bmatrix} 1 & 2 & 3 \\ 4 & 5 & 6 \end{bmatrix}$; (b) $\begin{bmatrix} 1 \\ 2 \\ 3 \end{bmatrix}$; (c) $[1 \ 2 \ 3]$; (d) $\begin{bmatrix} 1 & 2 \\ 3 & 4 \\ 5 & 6 \end{bmatrix}$.

Options The matrix is associated with a linear transformation:

(i) $\mathbb{R}^1 \to \mathbb{R}^3$, (ii) $\mathbb{R}^3 \to \mathbb{R}^2$, (iii) $\mathbb{R}^3 \to \mathbb{R}^1$, (iv) $\mathbb{R}^2 \to \mathbb{R}^3$.

3 For each of the following matrices A, vector spaces V, W, write down the linear transformation associated with A with respect to the given bases.

(a) $A = \begin{bmatrix} 1 & 2 & 3 \\ 4 & 5 & 6 \end{bmatrix}$; $V = \mathbb{R}^3$, $W = \mathbb{R}^2$; standard bases for $\mathbb{R}^3, \mathbb{R}^2$;

(b) $A = \begin{bmatrix} 1 & 2 \\ 3 & 4 \end{bmatrix}$; $V = \mathbb{R}^2$, $W = \mathbb{C}$; standard basis for $\mathbb{R}^2$, basis $1, i$ for $\mathbb{C}$;

(c) $A = \begin{bmatrix} 1 & -1 & 0 \\ 2 & 3 & 1 \\ 1 & 4 & 5 \end{bmatrix}$; $V = P_2$, $W = \mathbb{R}^3$; basis $1, x, x^2$ for P_2, basis $(1,0,0), (1,-1,0), (-1,-1,-1)$ for $\mathbb{R}^3$.

4 Prove parts (a) and (b) of Theorem 6.3.1.

5 Justify the assertion made in the text that, given a matrix A and bases for V, W, the linear transformation from V to W defined by the equations (1) (p. 131) is unique.

6.4 KERNEL AND IMAGE OF A LINEAR TRANSFORMATION

There are two subspaces which may be associated with any linear transformation and which play an important role in the study of that transformation. Let V, W be finite-dimensional vector spaces over the same field F, and let $T: V \to W$ be a linear transformation. The *kernel* and *image* of T, denoted by ker T and im T respectively, are defined by

$$\ker T = \{v \in V : T(v) = \mathbf{0}\},$$

$$\operatorname{im} T = \{w \in W : w = T(v) \text{ for some } v \in V\}.$$

It is clear from the definitions that ker T is a subset of V, and that im T is a subset of W (Fig. 6.2).

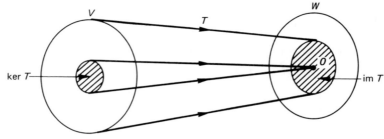

Fig. 6.2

THEOREM 6.4.1

(a) ker T is a subspace of V; and
(b) im T is a subspace of W.

Proof

(a) We saw in Lemma 6.1.1 that $T(\mathbf{0}) = \mathbf{0}$, so that $\mathbf{0} \in \ker T$, and ker T is non-empty. Hence S1 is satisfied. Suppose that $\mathbf{v}, \mathbf{v}_1, \mathbf{v}_2 \in \ker T$. Then $T(\mathbf{v}) = T(\mathbf{v}_1) = T(\mathbf{v}_2) = \mathbf{0}$, and

$$T(\mathbf{v}_1 + \mathbf{v}_2) = T(\mathbf{v}_1) + T(\mathbf{v}_2) = \mathbf{0} + \mathbf{0} = \mathbf{0},$$

$$T(\alpha \mathbf{v}) = \alpha T(\mathbf{v}) = \alpha \mathbf{0} = \mathbf{0} \quad (\alpha \in F).$$

Thus, $\mathbf{v}_1 + \mathbf{v}_2, \alpha \mathbf{v} \in \ker T$ and S2 and S3 are satisfied.

(b) The fact that $T(\mathbf{0}) = \mathbf{0}$ also tells us that $\mathbf{0} \in \operatorname{im} T$. Suppose that $\mathbf{w}, \mathbf{w}_1, \mathbf{w}_2 \in \operatorname{im} T$. Then $\mathbf{w} = T(\mathbf{v})$, $\mathbf{w}_1 = T(\mathbf{v}_1)$, $\mathbf{w}_2 = T(\mathbf{v}_2)$ for some $\mathbf{v}, \mathbf{v}_1, \mathbf{v}_2 \in V$, and

$$\mathbf{w}_1 + \mathbf{w}_2 = T(\mathbf{v}_1) + T(\mathbf{v}_2) = T(\mathbf{v}_1 + \mathbf{v}_2) \in \operatorname{im} T,$$

$$\alpha \mathbf{w} = \alpha T(\mathbf{v}) = T(\alpha \mathbf{v}) \in \operatorname{im} T.$$

It therefore follows again that S1, S2 and S3 hold.

Example 6.4.1

Let $T: \mathbb{R}^3 \to \mathbb{R}^2$ be defined by $T((x, y, z)) = (x, x)$. Then

$$\begin{aligned}
\operatorname{im} T &= \{(a, b) \in \mathbb{R}^2 : (a, b) = T((x, y, z)) \quad \text{for some } x, y, z \in \mathbb{R}\} \\
&= \{(a, b) \in \mathbb{R}^2 : (a, b) = (x, x) \quad \text{for some } x \in \mathbb{R}\} \\
&= \{(x, x) : x \in \mathbb{R}\} \\
&= \{x(1, 1) : x \in \mathbb{R}\} \\
&= \langle (1, 1) \rangle.
\end{aligned}$$

Also, $\ker T = \{(x, y, z) \in \mathbb{R}^3 : T((x, y, z)) = (0, 0)\}$
$= \{(x, y, z) \in \mathbb{R}^3 : (x, x) = (0, 0)\}$
$= \{(0, y, z) : y, z \in \mathbb{R}\}$
$= \{(0, y, 0) + (0, 0, z) : y, z \in \mathbb{R}\}$
$= \{y(0, 1, 0) + z(0, 0, 1) : y, z \in \mathbb{R}\}.$

It is clear then that $(0, 1, 0), (0, 0, 1)$ span $\ker T$. But it is easily checked that they are also linearly independent, so
$$\ker T = \langle (0, 1, 0), (0, 0, 1) \rangle,$$
$\dim(\operatorname{im} T) = 1$, $\dim(\ker T) = 2$ and
$$\dim(\operatorname{im} T) + \dim(\ker T) = 1 + 2 = 3 = \dim \mathbb{R}^3 = \dim(\text{domain space}).$$

This is an important observation, and is true generally, as is shown below.

THEOREM 6.4.2 (The Dimension Theorem) Let $T: V \to W$ be a linear transformation. Then
$$\dim(\operatorname{im} T) + \dim(\ker T) = \dim V.$$

Proof Let $\mathbf{v}_1, \ldots, \mathbf{v}_s$ be a basis for $\ker T$. Then, by Theorem 5.4.4, we can extend to a basis $\mathbf{v}_1, \ldots, \mathbf{v}_m$ for V, where $m = \dim V$ and $m \geq s$. If $m = s$, then $\ker T = V$, $\operatorname{im} T = \{\mathbf{0}\}$, and the result is clear. So suppose that $m > s$. We show that $T(\mathbf{v}_{s+1}), \ldots, T(\mathbf{v}_m)$ is a basis for $\operatorname{im} T$. This suffices, since then $\dim(\operatorname{im} T) + \dim(\ker T) = m = \dim V$.

Linear independence Suppose that
$$\mathbf{0} = \alpha_{s+1} T(\mathbf{v}_{s+1}) + \ldots + \alpha_m T(\mathbf{v}_m) = T(\alpha_{s+1} \mathbf{v}_{s+1} + \ldots + \alpha_m \mathbf{v}_m).$$
Then $\alpha_{s+1} \mathbf{v}_{s+1} + \ldots + \alpha_m \mathbf{v}_m \in \ker T$, and so
$$\alpha_{s+1} \mathbf{v}_{s+1} + \ldots + \alpha_m \mathbf{v}_m = \alpha_1 \mathbf{v}_1 + \ldots + \alpha_s \mathbf{v}_s \quad \text{for some } \alpha_1, \ldots, \alpha_s \in F.$$
Rearranging this equation gives
$$\alpha_1 \mathbf{v}_1 + \ldots + \alpha_s \mathbf{v}_s - \alpha_{s+1} \mathbf{v}_{s+1} - \ldots - \alpha_m \mathbf{v}_m = \mathbf{0};$$
whence $\alpha_1 = \ldots = \alpha_s = \alpha_{s+1} = \ldots = \alpha_m = 0$ because $\mathbf{v}_1, \ldots, \mathbf{v}_m$ are linearly independent. We conclude that $T(\mathbf{v}_{s+1}), \ldots, T(\mathbf{v}_m)$ are linearly independent.

Spanning Let $\mathbf{w} \in \operatorname{im} T$. Then $\mathbf{w} = T(\mathbf{v})$ for some $\mathbf{v} \in V$. Now
$$\mathbf{v} = \lambda_1 \mathbf{v}_1 + \ldots + \lambda_s \mathbf{v}_s + \lambda_{s+1} \mathbf{v}_{s+1} + \ldots + \lambda_m \mathbf{v}_m \quad \text{for some } \lambda_1, \ldots, \lambda_m \in F.$$

Thus,

$$\mathbf{w} = \lambda_1 T(\mathbf{v}_1) + \ldots + \lambda_s T(\mathbf{v}_s) + \lambda_{s+1} T(\mathbf{v}_{s+1}) + \ldots + \lambda_m T(\mathbf{v}_m)$$
$$= \lambda_{s+1} T(\mathbf{v}_{s+1}) + \ldots + \lambda_m T(\mathbf{v}_m),$$

since $\mathbf{v}_1, \ldots, \mathbf{v}_s \in \ker T$, and so $T(\mathbf{v}_1) = \ldots = T(\mathbf{v}_s) = \mathbf{0}$. It follows that $T(\mathbf{v}_{s+1}), \ldots, T(\mathbf{v}_m)$ span im T, and hence that they are a basis for im T.

The dimension of the image space of T is often called the *rank* of T; likewise, the dimension of the kernel is often referred to as the *nullity* of T. Theorem 6.4.2 then reads

$$\operatorname{rank} T + \operatorname{nullity} T = \dim V.$$

The dimension theorem is an important theoretical result. In particular, in conjunction with the next theorem, it helps us to calculate the image of a linear transformation.

THEOREM 6.4.3   Let $T: V \to W$ be a linear transformation, and let $\mathbf{e}_1, \ldots, \mathbf{e}_n$ be a basis for V. Then $T(\mathbf{e}_1), \ldots, T(\mathbf{e}_n)$ span im T.

Proof   This is quite easy and will be set as an exercise.

Example 6.4.2

Let $T: \mathbb{R}^3 \to \mathbb{R}^3$ be defined by $T((x, y, z)) = (x + y - z, 3x - y - z, 5x + y - 3z)$. Then

$$(x, y, z) \in \ker T \Leftrightarrow (0, 0, 0) = T((x, y, z)) = (x + y - z, 3x - y - z, 5x + y - 3z)$$
$$\Leftrightarrow x + y - z = 0,$$
$$3x - y - z = 0,$$
$$5x + y - 3z = 0.$$

By way of revision we will employ the row-reduction techniques of section 3.2 to solve this system of equations.

$$\begin{bmatrix} 1 & 1 & -1 \\ 3 & -1 & -1 \\ 5 & 1 & -3 \end{bmatrix} \to \begin{bmatrix} 1 & 1 & -1 \\ 0 & -4 & 2 \\ 0 & -4 & 2 \end{bmatrix} \quad \begin{array}{l} R_2 = r_2 - 3r_1, \\ R_3 = r_3 - 5r_1. \end{array}$$

(Note that the final column of zeroes can be omitted here, as it does not change under any elementary row operation.)

An equivalent system is, therefore,
$$x + y - z = 0$$
$$-4y + 2z = 0.$$
This has infinitely many solutions: $2y = z$, $x = z - y = 2y - y = y$. Hence,
$$\ker T = \{(y, y, 2y) : y \in \mathbb{R}\}$$
$$= \{y(1, 1, 2) : y \in \mathbb{R}\} = \langle (1, 1, 2) \rangle,$$
and nullity $T = 1$. Now $T((1, 0, 0)) = (1, 3, 5)$, $T((0, 1, 0)) = (1, -1, 1)$, $T((0, 0, 1)) = (-1, -1, -3)$. But $(1, 0, 0)$, $(0, 1, 0)$, $(0, 0, 1)$ is a basis for $\mathbb{R}^3$, so $(1, 3, 5)$, $(1, -1, 1)$, $(-1, -1, -3)$ is a spanning sequence for im T, by Theorem 6.4.3. These vectors must be linearly dependent, since, by the dimension theorem,
$$\text{rank } T = \dim \mathbb{R}^3 - \text{nullity } T = 3 - 1 = 2.$$
In fact, it is easy to check that
$$(1, 3, 5) = -1(1, -1, 1) - 2(-1, -1, -3).$$
In order to obtain a basis for im T we have simply to choose a linearly independent subsequence of this spanning sequence with two vectors in it.

Hence
$$\text{im } T = \langle (1, -1, 1), (-1, -1, -3) \rangle$$
$$= \langle (1, -1, 1), (1, 3, 5) \rangle$$
$$= \langle (1, 3, 5), (-1, -1, -3) \rangle.$$

Finally we show how the kernel and the image can be used to determine whether T is injective or surjective.

THEOREM 6.4.4 Let $T : V \to W$ be a linear transformation. Then the following are equivalent:

(a) T is surjective;   (b) im $T = W$;   (c) rank $T = \dim W$;
(d) nullity $T = \dim V - \dim W$.

Proof This is not difficult, and will be set as an exercise.

THEOREM 6.4.5 Let $T : V \to W$ be a linear transformation. Then the following are equivalent:

(a) T is injective;   (b) ker $T = \{\mathbf{0}\}$;   (c) nullity $T = 0$;
(d) rank $T = \dim V$.

Proof The equivalence of (b) and (c) is clear; the equivalence of (c) and (d) follows from the dimension theorem. So it is sufficient to show that (a) is equivalent to (b).

(a)$\Rightarrow$(b) Suppose that T is injective. Then
$$\mathbf{v} \in \ker T \Leftrightarrow T(\mathbf{v}) = \mathbf{0} = T(\mathbf{0}) \Leftrightarrow \mathbf{v} = \mathbf{0}.$$

Hence $\ker T = \{\mathbf{0}\}$.

(b)$\Rightarrow$(a) Suppose that $\ker T = \{\mathbf{0}\}$. Then
$$T(\mathbf{x}) = T(\mathbf{y}) \Rightarrow T(\mathbf{x} - \mathbf{y}) = \mathbf{0} \Rightarrow \mathbf{x} - \mathbf{y} \in \ker T \Rightarrow \mathbf{x} - \mathbf{y} = \mathbf{0} \Rightarrow \mathbf{x} = \mathbf{y},$$

so T is injective.

Example 6.4.3

Show that the linear transformation $T: \mathbb{R}^2 \to \mathbb{R}^2$ given by
$$T((x, y)) = (x + y, x - y)$$
is bijective.

Solution $(x, y) \in \ker T \Leftrightarrow x + y = 0, x - y = 0 \Leftrightarrow x = y = 0$, so $\ker T = \{\mathbf{0}\}$. Thus, T is injective, by Theorem 6.4.5. Also, nullity $T = 0 = 2 - 2 = \dim \mathbb{R}^2 - \dim \mathbb{R}^2$, so T is surjective, by Theorem 6.4.4.

EXERCISES 6.4

1 For each of the following linear transformations T, find im T and ker T.
  (a) $T: \mathbb{R}^2 \to \mathbb{R}^2$ given by $T((x, y)) = (x + y, 0)$;
  (b) $T: \mathbb{R}^2 \to \mathbb{R}^2$ given by $T((x, y)) = (x + y, x - y)$;
  (c) $T: \mathbb{R}^3 \to \mathbb{R}^3$ given by $T((x, y, z)) = (x, y, y)$;
  (d) $T: \mathbb{R}^3 \to \mathbb{R}^3$ given by
  $$T((x, y, z)) = (2x - y + z, -x + 3y + 5z, 10x - 9y - 7z);$$
  (e) $T: M_{2,2} \to \mathbb{R}^3$ given by $T\left(\begin{bmatrix} a & b \\ c & d \end{bmatrix}\right) = (a + b - c, b + d, a - c - d)$;
  (f) $T: P_2 \to P_2$ given by $T(a_0 + a_1 x + a_2 x^2) = a_1 + 2a_2 x$.

2 Prove Theorem 6.4.3.

3 Prove Theorem 6.4.4.

4 Which of the following linear transformations are (i) surjective, (ii) injective, (iii) bijective?

  (a) $T: \mathbb{R}^2 \to \mathbb{R}^2$ given by $T((x, y)) = (3x + 2y, x - 3y)$;
  (b) $T: \mathbb{R}^3 \to \mathbb{R}^2$ given by $T((x, y, z)) = (x - y, x + z)$;

(c) $T: \mathbb{R}^2 \to \mathbb{R}^2$ given by $T((x,y)) = (x+y, 3x+3y)$;
(d) $T: \mathbb{R}^2 \to \mathbb{R}^3$ given by $T((x,y)) = (x, y, y)$.

5 Let V, W be vector spaces of the same finite dimension, and let $T: V \to W$ be a linear transformation. Prove that T is injective if and only if T is surjective.

6 Let $T: U \to V$, $S: V \to W$ be linear transformations. Prove that
   (a) $\operatorname{rank}(S \circ T) \leqslant \operatorname{rank} S$;
   (b) $\operatorname{rank}(S \circ T) = \operatorname{rank} T$ if S is injective;
   (c) $\operatorname{rank}(S \circ T) = \operatorname{rank} S$ if T is surjective.

6.5 THE RANK OF A MATRIX

Let $A = [a_{ij}]$ be an $m \times n$ matrix. In Chapter 2 we allowed the elements of a matrix to be only real or complex numbers. However, we could have allowed them to belong to any field F, and all the results we proved would have remained valid; in fact, the same proof would have worked in every case apart from Corollary 4.3.7. So, here we suppose only that $a_{ij} \in F$, a field.

Let $\mathbf{e}_i$ be the vector with n coordinates all of which are zero apart from the ith, which is 1; likewise, allow $\mathbf{f}_j$ to be the vector with m coordinates all of which are zero apart from the jth, which is 1. Thus,

$$\mathbf{e}_i = (0, \ldots, 0, \underset{\uparrow}{1}, 0, \ldots, 0)$$
$$i\text{th position}$$

with n entries, and

$$\mathbf{f}_j = (0, \ldots, 0, \underset{\uparrow}{1}, 0, \ldots, 0)$$
$$j\text{th position}$$

with m entries. Then $\mathbf{e}_1, \ldots, \mathbf{e}_n$ is a basis for F^n, and $\mathbf{f}_1, \ldots, \mathbf{f}_m$ is a basis for F^m; we call these the *standard bases*. Let $T: F^n \to F^m$ be the linear transformation corresponding to A with respect to these bases, so that

$$T(\mathbf{e}_i) = a_{1i}\mathbf{f}_1 + \ldots + a_{mi}\mathbf{f}_m \quad (1 \leqslant i \leqslant n).$$

We define the *rank* of A to be the rank of T.

Example 6.5.1

Let $A = \begin{bmatrix} 1 & -1 & 2 \\ 2 & 1 & -1 \\ 1 & -4 & 7 \end{bmatrix}$. Then the associated linear transformation referred to above is $T: \mathbb{R}^3 \to \mathbb{R}^3$ given by

$$T((x, y, z)) = (x - y + 2z, 2x + y - z, x - 4y + 7z).$$

Now $T((1, 0, 0)) = (1, 2, 1)$, $T((0, 1, 0)) = (-1, 1, -4)$, $T((0, 0, 1)) = (2, -1, 7)$ span im T, by Theorem 6.4.3. However,

$$(1, 2, 1) = 5(-1, 1, -4) + 3(2, -1, 7),$$

so these three vectors are linearly dependent. Since $(1, 2, 1)$, $(-1, 1, -4)$ are linearly independent, these two vectors form a basis for im T. Hence rank A = rank T = 2.

The *row space* of the matrix A is the subspace of F^m spanned by the rows of A; its dimension is the *row rank* of A. In similar fashion the *column space* of A is the subspace of F^n spanned by the columns of A; its dimension is the *column rank* of A.

Let us examine the above example again. There the

column space of $A = \langle (1, 2, 1), (-1, 1, -4), (2, -1, 7) \rangle$,

which we saw has a basis comprising $(1, 2, 1), (-1, 1, -4)$. Thus, the column rank of $A = 2 =$ the rank of A. These observations suggest the next result.

LEMMA 6.5.1   rank A = column rank A.

Proof Let $\mathbf{e}_1, \ldots, \mathbf{e}_n$ and $\mathbf{f}_1, \ldots, \mathbf{f}_m$ be the standard bases for F^n and F^m respectively, and let T be the linear transformation corresponding to A. Then

$$T(\mathbf{e}_i) = a_{1i}\mathbf{f}_1 + \ldots + a_{mi}\mathbf{f}_m \qquad (1 \leqslant i \leqslant n)$$
$$= (a_{1i}, \ldots, a_{mi}),$$

which is the ith column of A. By definition,

rank A = rank T = dim(im T).

But im T is spanned by $T(\mathbf{e}_1), \ldots, T(\mathbf{e}_n)$; that is, im T is spanned by the columns of A. Thus, im $T =$ column space of A, and the result follows.

This lemma prompts the thought of whether rank A is also equal to the row rank of A. This question is a little harder to settle, but is resolved by the theorem below.

THEOREM 6.5.2   rank A = column rank A = row rank A.

LINEAR TRANSFORMATIONS

Proof We adopt the same notation as in Lemma 6.5.1 above, and let row rank $A = r$.

Then
$$\ker T = \{\mathbf{x} \in F^n : T(\mathbf{x}) = \mathbf{0}\}$$
$$= \{\mathbf{x} \in F^n : A\mathbf{x}^T = O\} \quad \text{(see section 6.3)}.$$

Now the matrix equation $A\mathbf{x}^T = O$ corresponds to the system of linear homogeneous equations
$$a_{i1}x_1 + \ldots + a_{in}x_n = 0 \quad (1 \leq i \leq m).$$

Row-reduce the matrix A to its reduced echelon form, B. Then
$$r = \text{row rank of } A = \text{row rank of } B.$$

Also, $\dim(\text{solution space of } B\mathbf{x}^T = O) = \dim(\text{solution space of } A\mathbf{x}^T = O)$
$$= \dim(\ker T)$$
$$= n - \dim(\text{im } T)$$
$$= n - \text{rank } A.$$

Now $\dim(\text{solution space of } B\mathbf{x}^T = O) = \dim(\ker S)$, where
$$S : F^n \to F^r : \mathbf{x} \mapsto B\mathbf{x}^T.$$

Clearly $\text{im } S \subset F^r$, so $\dim(\text{im } S) \leq r$, which implies that
$$\dim(\ker S) = n - \dim(\text{im } S) \geq n - r.$$

Hence
$$n - \text{rank } A = \dim(\text{solution space of } B\mathbf{x}^T = O) = \dim(\ker S) \geq n - r,$$

from which rank $A \leq r$. It follows that column rank $A \leq$ row rank A. Finally, applying this result to A^T gives the reverse inequality.

You may have to read the above proof several times in order to digest it. If you have difficulties, do not be discouraged: concentrate on understanding what the result itself says and on being able to use it. You can come back and read the proof again later.

Example 6.5.2

Let A be as in example 6.5.1 above. Then
$$\text{row space } A = \langle (1, -1, 2), (2, 1, -1), (1, -4, 7) \rangle$$
$$= \langle (1, -1, 2), (2, 1, -1) \rangle$$

since $(1, -4, 7) = 3(1, -1, 2) - (2, 1, -1)$. Moreover, $(1, -1, 2), (2, 1, -1)$ are linearly independent, and so form a basis for the row space of A. Hence

$$\text{row rank } A = 2 = \text{column rank } A = \text{rank } A.$$

We close this section by collecting together some properties of the rank of a matrix.

THEOREM 6.5.3 Let $A = [a_{ij}]_{m \times n}$, $B = [b_{ij}]_{n \times n}$ be matrices over the same field F. Then

(a) rank A^T = rank A;
(b) rank(EA) = rank A for any $m \times m$ elementary matrix E;
(c) rank(AE) = rank A for any $n \times n$ elementary matrix E;
(d) rank(AB) = rank A if B is non-singular;
(e) rank $B = n \Leftrightarrow B$ is non-singular;
(f) rank$(AB) \leq$ rank A.

Proof

(a) This is immediate from Theorem 6.5.2.
(b) Premultiplying A by E is equivalent to performing the corresponding e.r.o. on A (see section 3.5). But e.r.o.s do not change the row rank of A. (This should be checked, and is set as exercise 6.5.3.) The result thus follows from Theorem 6.5.2.
(c) This is similar to (b).
(d) Since B is non-singular, it can be written as $B = E_1 \ldots E_r$, where $E_1, \ldots, E_r$ are elementary matrices (see the first paragraph of section 3.6). The result is now seen to be a consequence of (c).
(e) Here we need apply only Theorems 6.4.4 and 6.4.5.
(f) This is set as exercise 6.5.6.

COROLLARY 6.5.4 Row-equivalent matrices have the same rank. In particular, the rank of a matrix is equal to the number of non-zero rows in the reduced echelon form of that matrix.

Proof The first assertion follows from Theorem 6.5.3 and Corollary 3.5.2. (Actually, we require a modified form of (d), namely that rank (BA) = rank A, when the product BA is defined, but this is easily checked, and is set as exercise 6.5.5.)

Now if you look back to section 3.4 it can be seen from the definition that the row rank of a matrix is equal to the number of non-zero rows in its reduced echelon form.

Example 6.5.3

Let A be as in example 6.5.1 yet again. Row reducing A gives

$$\begin{bmatrix} 1 & -1 & 2 \\ 2 & 1 & -1 \\ 1 & -4 & 7 \end{bmatrix}$$
$$\downarrow$$
$$\begin{bmatrix} 1 & -1 & 2 \\ 0 & 3 & -5 \\ 0 & -3 & 5 \end{bmatrix} \begin{matrix} R_2 = r_2 - 2r_1 \\ R_3 = r_3 - r_1 \end{matrix}$$
$$\downarrow$$
$$\begin{bmatrix} 1 & -1 & 2 \\ 0 & 3 & -5 \\ 0 & 0 & 0 \end{bmatrix} R_3 = r_3 + r_2$$
$$\downarrow$$
$$\begin{bmatrix} 1 & -1 & 2 \\ 0 & 1 & -\frac{5}{3} \\ 0 & 0 & 0 \end{bmatrix} R_2 = \tfrac{1}{3} r_2$$
$$\downarrow$$
$$\begin{bmatrix} 1 & 0 & \frac{1}{3} \\ 0 & 1 & -\frac{5}{3} \\ 0 & 0 & 0 \end{bmatrix} R_1 = r_1 + r_2$$

Hence rank $A = 2$.

Note: The astute reader will have notice that the last two steps in the above reduction were unnecessary. This suggests a strengthening of Corollary 6.5.4. Can the reader see what this more powerful version might be? The answer appears in the exercises.

EXERCISES 6.5

1 Determine the rank of each of the following matrices:

(a) $\begin{bmatrix} 1 & -1 \\ 3 & 5 \end{bmatrix}$; (b) $\begin{bmatrix} -1 & -2 & 1 \\ 2 & 1 & 2 \\ 5 & 2 & 2 \end{bmatrix}$; (c) $\begin{bmatrix} 4 & 3 & 0 & 1 \\ 3 & 2 & 1 & 1 \\ 7 & 5 & 1 & 2 \\ 6 & 5 & -2 & 1 \end{bmatrix}$;

(d) $\begin{bmatrix} 3 & 5 & -2 & 2 \\ 1 & 2 & -1 & 1 \\ 1 & 2 & -2 & 1 \end{bmatrix}.$

2 Write down the linear transformation (as defined at the beginning of this section) corresponding to each of the above matrices, and in each case find a basis for the image space.

3 Prove that e.r.o.s do not change the row rank of a matrix.

4 Write out a full proof for part (c) of Theorem 6.5.3.

5 Let B be an $m \times m$ matrix, and A an $m \times n$ matrix. Prove that rank(BA) = rank A if B is non-singular.

6 Prove part (f) of Theorem 6.5.3.

7 Prove that the rank of a matrix is equal to the number of non-zero rows in the echelon form of that matrix.

8 Interpret Theorem 6.5.3 (d) in terms of linear transformations, and use this interpretation to give a different proof of this result.

6.6 SYSTEMS OF LINEAR EQUATIONS

Consider the following system of linear equations again:

$$a_{11}x_1 + a_{12}x_2 + \ldots + a_{1n}x_n = b_1$$
$$a_{21}x_1 + a_{22}x_2 + \ldots + a_{2n}x_n = b_2$$
$$\ldots\ldots\ldots\ldots\ldots\ldots\ldots\ldots\ldots\ldots\ldots$$
$$a_{m1}x_1 + a_{m2}x_2 + \ldots + a_{mn}x_n = b_m.$$

We have seen that this system can be written in matrix form as $AX = B$, where

$$A = \begin{bmatrix} a_{11} & a_{12} & \ldots & a_{1n} \\ a_{21} & a_{22} & \ldots & a_{2n} \\ \ldots & \ldots & \ldots & \ldots \\ a_{m1} & a_{m2} & \ldots & a_{mn} \end{bmatrix}, \quad X = \begin{bmatrix} x_1 \\ x_2 \\ \vdots \\ x_n \end{bmatrix}, \quad B = \begin{bmatrix} b_1 \\ b_2 \\ \vdots \\ b_m \end{bmatrix}.$$

Armed with the concept of a linear transformation, we can now view the system in yet another way. If $T: F^n \to F^m$ is the linear transformation associated with A with respect to the standard bases in F^m, F^n, then we may regard the problem of solving the system as being equivalent to that of finding those vectors $\mathbf{x} \in F^n$ for which $T(\mathbf{x}) = \mathbf{b}$, where $\mathbf{x} = (x_1, \ldots, x_n)(= X^T)$, $\mathbf{b} = (b_1, \ldots, b_m)(= B^T)$. We will utilise these ideas in order to gain more insight into the solutions to this system.

LINEAR TRANSFORMATIONS

Throughout we will fix $T, A, X, B, \mathbf{x}, \mathbf{b}$ to be as described above. We will also denote the matrix obtained from A by including an $(n+1)$th column consisting of B by $(A|B)$; that is,

$$(A|B) = \begin{bmatrix} a_{11} & \ldots & a_{1n} & b_1 \\ a_{21} & \ldots & a_{2n} & b_2 \\ \multicolumn{4}{c}{\dotfill} \\ a_{m1} & \ldots & a_{mn} & b_m \end{bmatrix}$$

THEOREM 6.6.1 The following conditions are equivalent:
(a) the equations $AX = B$ are consistent;
(b) $\mathbf{b} \in \operatorname{im} T$;
(c) $\mathbf{b} \in$ column space of A;
(d) rank A = rank $(A|B)$.

Proof

(a)$\Rightarrow$(b) It follows from the definition of T that if X is a solution of the system then $T(\mathbf{x}) = \mathbf{b}$ (where $\mathbf{x} = X^T$). Thus $\mathbf{b} \in \operatorname{im} T$.

(b)$\Rightarrow$(c) Let $\mathbf{e}_1, \ldots, \mathbf{e}_n$ and $\mathbf{f}_1, \ldots, \mathbf{f}_m$ be the standard bases for F^n and F^m respectively. Suppose that $\mathbf{b} \in \operatorname{im} T$. Then $\mathbf{b} = T(\mathbf{x})$ for some $\mathbf{x} \in F^n$, and $\mathbf{x} = (x_1, \ldots, x_n) = x_1 \mathbf{e}_1 + \ldots + x_n \mathbf{e}_n$. Hence

$$\mathbf{b} = T(\mathbf{x}) = x_1 T(\mathbf{e}_1) + \ldots + x_n T(\mathbf{e}_n)$$
$$= x_1(a_{11}, \ldots, a_{m1}) + \ldots + x_n(a_{1n}, \ldots, a_{mn}) \in \text{column space of } A$$

(as in Lemma 6.5.1).

(c)$\Rightarrow$(d) Let $\mathbf{b} \in$ column space of A. Then B is a linear combination of the columns of A. Hence

column space of A = column space of $(A|B)$.

Thus, rank A = rank $(A|B)$, by Lemma 6.5.1.

(d)$\Rightarrow$(a) Assume that rank A = rank $(A|B)$. Then, since the column space of A is contained in the column space of $(A|B)$ and these two subspaces have the same dimension,

column space of A = column space of $(A|B)$.

Thus, B is a linear combination of the columns of A; that is,

$$\begin{bmatrix} b_1 \\ \vdots \\ b_m \end{bmatrix} = \lambda_1 \begin{bmatrix} a_{11} \\ \vdots \\ a_{m1} \end{bmatrix} + \ldots + \lambda_n \begin{bmatrix} a_{1n} \\ \vdots \\ a_{mn} \end{bmatrix} = \begin{bmatrix} a_{11}\lambda_1 + \ldots + a_{1n}\lambda_n \\ \dotfill \\ a_{m1}\lambda_1 + \ldots + a_{mn}\lambda_n \end{bmatrix}.$$

It is clear that $x_i = \lambda_i$ is a solution to the system $AX = B$.

Example 6.6.1

Determine whether the system $\quad x + y = 0 \quad$ is consistent.
$$x - y + z = 1$$
$$2x - y - z = 1$$

Solution Let $A = \begin{bmatrix} 1 & 1 & 0 \\ 1 & -1 & 1 \\ 2 & -1 & -1 \end{bmatrix}$. First we calculate the rank of A.

Suppose that $\quad \alpha(1, 1, 2) + \beta(1, -1, -1) + \gamma(0, 1, -1) = (0, 0, 0)$.
Then $\quad \alpha + \beta = 0, \alpha - \beta + \gamma = 0, 2\alpha - \beta - \gamma = 0$.
Solving these equations, we find that $\alpha = \beta = \gamma = 0$.
Hence, $(1, 1, 2), (1, -1, -1), (0, 1, -1)$ are linearly independent and so form a basis for the column space of A. It follows that rank $A = 3$.

Also, by Theorem 6.5.2, $3 \geqslant$ rank $(A|B)$ (as $(A|B)$ has only three rows). But, clearly, rank $(A|B) \geqslant$ rank $A = 3$, so rank $(A|B) = 3 =$ rank A. Theorem 6.6.1 now tells us that the system has a solution.

Finally, we consider how to determine the number of solutions possessed by a given consistent system.

THEOREM 6.6.2 The set of solutions for the system $AX = 0$, considered as n-tuples, forms a subspace of F^n of dimension $n - r$, where $r =$ rank A.

Proof The set of solutions is
$$\{X : AX = 0\} = \{\mathbf{x}^T : T(\mathbf{x}) = \mathbf{0}\} = \{\mathbf{x}^T : \mathbf{x} \in \ker T\}.$$

It is easy to check that the mapping $\mathbf{x}^T \mapsto \mathbf{x}$ is an invertible linear transformation from this set onto ker T: its dimension is, therefore, equal to the nullity of $T = n -$ rank $T = n -$ rank A.

THEOREM 6.6.3 Let $X = X_1$ be a solution to the system $AX = B$ and let $X = X_2$ be a solution to the system $AX = 0$. Then $X = X_1 + X_2$ is also a solution to $AX = B$. Furthermore, *every* solution to $AX = B$ is of this form.

Proof We have $AX_1 = B$ and $AX_2 = 0$, so
$$A(X_1 + X_2) = AX_1 + AX_2 = B + 0 = B.$$

This establishes the first of our claims.

Now suppose that $AX_3 = B$.
Then
$$A(X_3 - X_1) = AX_3 - AX_1 = B - B = 0,$$
and so
$$X_3 - X_1 = X_2,$$
say, where $AX_2 = 0$. Thus, $X_3 = X_1 + X_2$.

Note: There is no theorem quite like Theorem 6.6.2 for the non-homogeneous case: in this instance the set of solutions does not form a subspace, since it has no zero element. However, we do have the following result.

THEOREM 6.6.4 Let A be an $n \times n$ matrix. Then the following are equivalent.
  (a) The equations $AX = B$ have a unique solution;
  (b) rank $A = n$;
  (c) A is non-singular.

Proof

(a)$\Rightarrow$(b) Suppose that $AX = B$ has a unique solution. By Theorem 6.6.3 the homogeneous system $AX = 0$ has only the trivial solution $X = 0$. It follows from Theorem 6.6.2 that $n - \text{rank } A = 0$. Thus, rank $A = n$.

(b)$\Rightarrow$(c) rank $A = n \Rightarrow$ rank $T = n \Rightarrow$ nullity $T = 0$

$\Rightarrow T$ is bijective   (Theorems 6.4.4 and 6.4.5)

$\Rightarrow T$ is invertible

$\Rightarrow A$ is non-singular   (Theorem 6.3.1).

(c)$\Rightarrow$(a) A is non-singular $\Rightarrow A^{-1}$ exists $\Rightarrow X = A^{-1}B$ is the unique solution.

Let us summarise then our results. The system $AX = B$
  (a) is inconsistent $\Leftrightarrow$ rank $A \neq \text{rank}(A|B)$,
  (b) has a unique solution $\Leftrightarrow \text{rank}(A|B) = \text{rank } A = n \Rightarrow A$ is non-singular,
  (c) has infinitely many solutions $\Leftrightarrow \text{rank}(A|B) = \text{rank } A < n$.

Example 6.6.2

Let us apply the theory to the system
$$x_1 + x_2 + x_3 = 1$$
$$2x_1 + x_2 + x_3 = 1$$
$$3x_1 + x_2 + x_3 = 1.$$

(This system was solved as an example in section 3.2.)

This can be written as $AX = B$ where $A = \begin{bmatrix} 1 & 1 & 1 \\ 2 & 1 & 1 \\ 3 & 1 & 1 \end{bmatrix}$, $X = \begin{bmatrix} x_1 \\ x_2 \\ x_3 \end{bmatrix}$, $B = \begin{bmatrix} \end{bmatrix}$

The reduced echelon form of A is $E = \begin{bmatrix} 1 & 0 & 0 \\ 0 & 1 & 1 \\ 0 & 0 & 0 \end{bmatrix}$. Hence rank $A = 2$, so the solution space of $AX = 0$ forms a subspace of $\mathbb{R}^3$ of dimension 1, by Theorem 6.6.2. In fact,

solution space of $AX \equiv 0 = $ solution space of $EX = 0$
$= \{(x_1, x_2, x_3)^T : x_1, x_2, x_3 \in \mathbb{R}, x_1 = 0, x_2 + x_3 = 0\}$
$= \{(0, x_2, -x_2)^T : x_2 \in \mathbb{R}\}$
$= \{x_2(0, 1, -1)^T : x_2 \in \mathbb{R}\}$
$= \langle (0, 1, -1)^T \rangle.$

It is clear that a particular solution to $AX = B$ is $x_1 = x_3 = 0$, $x_2 = 1$. It follows from Theorem 6.6.3 that the general solution to $AX = B$ is

$$(x_1, x_2, x_3) = (0, 1, 0) + \alpha(0, 1, -1);$$

i.e., $x_1 = 0$, $x_2 = 1 + \alpha$, $x_3 = -\alpha$ where $\alpha \in \mathbb{R}$, as found in section 3.2.

EXERCISES 6.6

1 Apply the methods of this section to exercises 3.2.

2 Show that the set of equations

$$x + y + z = 1$$
$$ax + ay + z = a + 1$$
$$ax + 2y + z = 2$$

has a unique solution for x, y, z if $a \neq 1, 2$. Discuss the cases $a = 1$ and $a = 2$ and, if solutions exist in these cases, find them.

3 Find all solutions of the system of linear equations

$$ax + 2y + z = 2a$$
$$2x + ay + z = -2$$
$$x + y + z = 1$$

for all values of a.

LINEAR TRANSFORMATIONS

4 Determine for what values of a the following system of equations is consistent:

$$(3+a)x + (2+2a)y + (a-2) = 0$$
$$(2a-3)x + (2-a)y + 3 = 0$$
$$3x + 7y - 1 = 0.$$

Solve the resulting equations for x, y in each case.

SOLUTIONS AND HINTS FOR EXERCISES

Exercises 6.1

1 (a) This is not linear. For example, $T((0,0)) = (0,1)$, which would contradict Lemma 6.1.1(a) if it were linear.

(b) This is linear. For,

$$T((x_1, y_1) + (x_2, y_2)) = T((x_1 + x_2, y_1 + y_2)) = (0, x_1 + x_2 + y_1 + y_2)$$
$$= (0, x_1 + y_1) + (0, x_2 + y_2)$$
$$= T((x_1, y_1)) + T((x_2, y_2)),$$

and $T(\alpha(x, y)) = T((\alpha x, \alpha y)) = (0, \alpha x + \alpha y) = \alpha(0, x + y) = \alpha T((x, y))$.

(c) This is not linear. For example, $T((1,0)) + T((1,0)) = (1,1) + (1,1) = (2,2)$, whereas $T((1,0) + (1,0)) = T((2,0)) = (4,2)$.

(d) Here $T((x, y)) = (y - 3x, x)$. This is linear. For,

$$T((x_1, y_1) + (x_2, y_2)) = T((x_1 + x_2, y_1 + y_2))$$
$$= (y_1 + y_2 - 3(x_1 + x_2), x_1 + x_2)$$
$$= (y_1 - 3x_1, x_1) + (y_2 - 3x_2, x_2)$$
$$= T((x_1, y_1)) + T((x_2, y_2)),$$

and

$$T(\alpha(x, y)) = T((\alpha x, \alpha y)) = (\alpha y - 3\alpha x, \alpha x) = \alpha(y - 3x, x) = \alpha T((x, y)).$$

2 (a) $(-1, 9)$; (b) $(0, 1)$; (c) $(4, 1)$; (d) $(-7, 2)$.

3 $T((a_1 + b_1 i) + (a_2 + b_2 i)) = T((a_1 + a_2) + (b_1 + b_2)i)$
$$= (a_1 + a_2) - (b_1 + b_2)i$$
$$= (a_1 - b_1 i) + (a_2 - b_2 i)$$
$$= T(a_1 + b_1 i) + T(a_2 + b_2 i),$$

and $T(\alpha(a + bi)) = T(\alpha a + \alpha bi) = \alpha a - \alpha bi = \alpha(a - bi) = \alpha T(a + bi).$

4 $T((a_1 + b_1 i) + (a_2 + b_2 i)) = T((a_1 + a_2) + (b_1 + b_2)i)$

$$= \begin{bmatrix} a_1 + a_2 & a_1 + a_2 + b_1 + b_2 \\ a_1 + a_2 - b_1 - b_2 & b_1 + b_2 \end{bmatrix}$$

$$= \begin{bmatrix} a_1 & a_1 + b_1 \\ a_1 - b_1 & b_1 \end{bmatrix} + \begin{bmatrix} a_2 & a_2 + b_2 \\ a_2 - b_2 & b_2 \end{bmatrix}$$

$$= T(a_1 + b_1 i) + T(a_2 + b_2 i),$$

and $\quad T(\alpha(a + bi)) = T(\alpha a + \alpha bi) = \begin{bmatrix} \alpha a & \alpha a + \alpha b \\ \alpha a - \alpha b & \alpha b \end{bmatrix}$

$$= \alpha \begin{bmatrix} a & a + b \\ a - b & b \end{bmatrix} = \alpha T(a + bi).$$

5 $T((a_1, a_2, a_3, \ldots) + (b_1, b_2, b_3, \ldots)) = T((a_1 + b_1, a_2 + b_2, a_3 + b_3, \ldots))$

$$= (0, a_1 + b_1, a_2 + b_2, \ldots)$$

$$= (0, a_1, a_2, \ldots) + (0, b_1, b_2, \ldots)$$

$$= T((a_1, a_2, a_3, \ldots)) + T((b_1, b_2, b_3, \ldots)),$$

and

$T(\alpha(a_1, a_2, a_3, \ldots)) = T((\alpha a_1, \alpha a_2, \alpha a_3, \ldots)) = (0, \alpha a_1, \alpha a_2, \ldots) = \alpha(0, a_1, a_2, \ldots)$

$$= \alpha T((a_1, a_2, a_3, \ldots)).$$

6 $(T_1 + T_2)(\mathbf{v}_1 + \mathbf{v}_2) = T_1(\mathbf{v}_1 + \mathbf{v}_2) + T_2(\mathbf{v}_1 + \mathbf{v}_2)$

$$= T_1(\mathbf{v}_1) + T_1(\mathbf{v}_2) + T_2(\mathbf{v}_1) + T_2(\mathbf{v}_2)$$

$$= (T_1 + T_2)(\mathbf{v}_1) + (T_1 + T_2)(\mathbf{v}_2),$$

and

$(T_1 + T_2)(\alpha \mathbf{v}) = T_1(\alpha \mathbf{v}) + T_2(\alpha \mathbf{v}) = \alpha T_1(\mathbf{v}) + \alpha T_2(\mathbf{v}) = \alpha(T_1(\mathbf{v}) + T_2(\mathbf{v}))$

$$= \alpha(T_1 + T_2)(\mathbf{v}), \text{ so } T_1 + T_2 \in L(V, W).$$

Also,

$\lambda T(\mathbf{v}_1 + \mathbf{v}_2) = \lambda(T(\mathbf{v}_1 + \mathbf{v}_2)) = \lambda(T(\mathbf{v}_1) + T(\mathbf{v}_2)) = \lambda T(\mathbf{v}_1) + \lambda T(\mathbf{v}_2),$

and $\quad \lambda T(\alpha \mathbf{v}) = \lambda(T(\alpha \mathbf{v})) = \lambda(\alpha T(\mathbf{v})) = \alpha(\lambda T)(\mathbf{v}), \text{ so } \lambda T \in L(V, W).$

7 Let $\mathbf{e}_1, \ldots, \mathbf{e}_n$ be a basis for V. Let $\mathbf{v} \in V$. There are *unique* elements $a_1, \ldots, a_n \in F$ such that $\mathbf{v} = a_1 \mathbf{e}_1 + \ldots + a_n \mathbf{e}_n$, and $T(\mathbf{v}) = T(a_1 \mathbf{e}_1 + \ldots + a_n \mathbf{e}_n) = a_1 T(\mathbf{e}_1) + \ldots + a_n T(\mathbf{e}_n)$, so the image of $T(\mathbf{v})$ is uniquely determined by $T(\mathbf{e}_1), \ldots, T(\mathbf{e}_n)$.

8 Let T be a linear transformation. Then
$$T(\alpha\mathbf{v} + \beta\mathbf{w}) = T(\alpha\mathbf{v}) + T(\beta\mathbf{w}) \quad \text{by T1}$$
$$= \alpha T(\mathbf{v}) + \beta T(\mathbf{w}) \quad \text{by T2.}$$
So suppose that $T(\alpha\mathbf{v} + \beta\mathbf{w}) = \alpha T(\mathbf{v}) + \beta T(\mathbf{w})$ for all $\alpha, \beta \in F$, and all $\mathbf{v}, \mathbf{w} \in V$. Putting $\alpha = \beta = 1$ we see that T1 holds; putting $\beta = 0$ shows that T2 holds.

9 Since U is a subspace, there is an element $\mathbf{u} \in U$. But then $T(\mathbf{u}) \in T(U)$, so S1 holds. Let $\mathbf{x}, \mathbf{y} \in T(U)$. Then $\mathbf{x} = T(\mathbf{u}), \mathbf{y} = T(\mathbf{u}')$ for some $\mathbf{u}, \mathbf{u}' \in U$, and
$$\mathbf{x} + \mathbf{y} = T(\mathbf{u}) + T(\mathbf{u}') = T(\mathbf{u} + \mathbf{u}') \in T(U),$$
$$\alpha\mathbf{x} = \alpha T(\mathbf{u}) = T(\alpha\mathbf{u}) \in T(U), \text{ so S2, S3 hold.}$$

Exercises 6.2

1 (a) Let $\quad T((x, y, z)) = (y - z, x + y, x + y + z) = (a, b, c)$.

Then $\quad a = y - z, b = x + y, c = x + y + z$,

which gives $\quad z = c - b, y = a + c - b, x = 2b - a - c$.

Hence $\quad T^{-1}((a, b, c)) = (2b - a - c, a + c - b, c - b)$.

(b) Let $\quad T(x + yi) = x - yi = a + bi$.

Then $\quad x = a, -y = b$,

which gives $\quad a = x, b = -y$.

Hence $\quad T^{-1}(a + bi) = a - bi, \quad$ and $\quad T^{-1} = T$.

(c) Let $\quad T(a_0 + a_1 x + a_2 x^2) = a_1 + a_2 x + a_0 x^2 = b_0 + b_1 x + b_2 x^2$.

Then $\quad a_1 = b_0, a_2 = b_1, a_0 = b_2$.

Hence $\quad T^{-1}(b_0 + b_1 x + b_2 x^2) = b_2 + b_0 x + b_1 x^2$.

2 The basis referred to is $T((1, 0, 0, 0)) = \begin{bmatrix} 1 & 0 \\ 0 & 0 \end{bmatrix}$, $T((1, 1, 0, 0)) = \begin{bmatrix} 1 & 1 \\ 0 & 0 \end{bmatrix}$,

$$T((1, 1, 1, 0)) = \begin{bmatrix} 1 & 1 \\ 1 & 0 \end{bmatrix}, T((1, 1, 1, 1)) = \begin{bmatrix} 1 & 1 \\ 1 & 1 \end{bmatrix}.$$

3 The linear transformation referred to is $T: \mathbb{R}^3 \to \mathbb{R}^3$ where

$$T((x, y, z)) = T(x(1, 0, 0) + y(0, 1, 0) + z(0, 0, 1))$$
$$= xT((1, 0, 0)) + yT((0, 1, 0)) + zT((0, 0, 1))$$
$$= x(1, 0, 0) + y(1, 1, 0) + z(1, 1, 1) = (x + y + z, y + z, z).$$

It is easy to check that $(1, 0, 0), (1, 1, 0), (1, 1, 1)$ is a basis for $\mathbb{R}^3$, and so T is invertible.

4 For example, $T: P_3 \to M_{2,2}$ given by

$$T(a_0 + a_1 x + a_2 x^2 + a_3 x^3) = \begin{bmatrix} a_0 & a_1 \\ a_2 & a_3 \end{bmatrix}$$

is an invertible transformation.

Exercises 6.3

1 (a) $\begin{bmatrix} -1 & 0 \\ 1 & 3 \\ 2 & -1 \end{bmatrix}$; (b) $\begin{bmatrix} 1 & 1 & 0 & 0 \\ 0 & 0 & 1 & 1 \end{bmatrix}$;

(c) $\begin{bmatrix} 0 & -\frac{1}{2} & -\frac{1}{2} \\ -1 & -\frac{3}{2} & -\frac{3}{2} \\ 0 & \frac{1}{2} & \frac{3}{2} \end{bmatrix}$; (d) $\begin{bmatrix} 1 & 0 & 0 & 1 \\ 0 & 1 & -1 & 0 \end{bmatrix}$.

2 (a) (ii); (b) (i); (c) (iii); (d) (iv).

3 (a) $T((x, y, z)) = (x + 2y + 3z, 4x + 5y + 6z)$;

(b) $T((x, y)) = (x + 2y) + (3x + 4y)i$;

(c) $\begin{bmatrix} 1 & -1 & 0 \\ 2 & 3 & 1 \\ 1 & 4 & 5 \end{bmatrix} \begin{bmatrix} a \\ b \\ c \end{bmatrix} = \begin{bmatrix} a - b \\ 2a + 3b + c \\ a + 4b + 5c \end{bmatrix}$ †

$= (a - b) \begin{bmatrix} 1 \\ 0 \\ 0 \end{bmatrix} + (2a + 3b + c) \begin{bmatrix} 1 \\ -1 \\ 0 \end{bmatrix}$

$+ (a + 4b + 5c) \begin{bmatrix} -1 \\ -1 \\ -1 \end{bmatrix} = \begin{bmatrix} 2a - 2b - 4c \\ -3a - 7b - 6c \\ -a - 4b - 5c \end{bmatrix}$

(*Note:* † indicates that these are coordinates with respect to a non-standard basis.)

Thus,

$$T(a + bx + cx^2) = (2a - 2b - 4c, -3a - 7b - 6c, -a - 4b - 5c).$$

4 Let the given bases be $\mathbf{e}_1, \ldots, \mathbf{e}_r$ for U, $\mathbf{f}_1, \ldots, \mathbf{f}_s$ for V, and suppose that $A = [a_{ij}]_{s \times r}$, $A_1 = [a'_{ij}]_{s \times r}$, $A_2 = [a''_{ij}]_{s \times r}$.

(a) $(T_1 + T_2)(\mathbf{e}_j) = T_1(\mathbf{e}_j) + T_2(\mathbf{e}_j) = \sum_{i=1}^{s} a'_{ij} \mathbf{f}_i + \sum_{i=1}^{s} a''_{ij} \mathbf{f}_i = \sum_{i=1}^{s} (a'_{ij} + a''_{ij}) \mathbf{f}_i.$

Hence, the matrix associated with $T_1 + T_2$ is $[a'_{ij} + a''_{ij}]_{s \times r} = A_1 + A_2$.

(b) $(\lambda T)(\mathbf{e}_j) = \lambda(T(\mathbf{e}_j)) = \lambda \sum_{i=1}^{s} a_{ij}\mathbf{f}_i = \sum_{i=1}^{s} (\lambda a_{ij})\mathbf{f}_i.$

Thus, the matrix associated with λT is $[\lambda a_{ij}]_{s \times r} = \lambda A$.

5 This follows from exercise 6.1.7.

Exercises 6.4

1 (a) $(x, y) \in \ker T \Leftrightarrow (x + y, 0) = (0, 0) \Leftrightarrow y = -x.$

Thus, $\ker T = \{(x, -x): x \in \mathbb{R}\} = \langle (1, -1) \rangle.$

Also, $\operatorname{im} T = \{(x + y, 0): x, y \in \mathbb{R}\} = \{(z, 0): z \in \mathbb{R}\} = \langle (1, 0) \rangle.$

(b) $(x, y) \in \ker T \Leftrightarrow (x + y, x - y) = (0, 0) \Leftrightarrow x + y = x - y = 0 \Leftrightarrow x = y = 0.$

Thus, $\ker T = \{(0, 0)\}$. It follows from the dimension theorem that $\operatorname{im} T = \mathbb{R}^2$.

(c) $(x, y, z) \in \ker T \Leftrightarrow (x, y, y) = (0, 0, 0) \Leftrightarrow x = y = 0.$

Thus $\ker T = \{(0, 0, z): z \in \mathbb{R}\} = \langle (0, 0, 1) \rangle.$

Also, $\operatorname{im} T = \{(x, y, y): x, y \in \mathbb{R}\} = \{x(1, 0, 0) + y(0, 1, 1): x, y \in \mathbb{R}\}$
$= \langle (1, 0, 0), (0, 1, 1) \rangle.$

(d) $(x, y, z) \in \ker T \Leftrightarrow (2x - y + z, -x + 3y + 5z, 10x - 9y - 7z) = (0, 0, 0)$

$\Leftrightarrow A\mathbf{x} = \mathbf{0}$, where $A = \begin{bmatrix} 2 & -1 & 1 \\ -1 & 3 & 5 \\ 10 & -9 & -7 \end{bmatrix}$, $\mathbf{x} = \begin{bmatrix} x \\ y \\ z \end{bmatrix}$.

The reduced echelon form of A is I_3, so the system $A\mathbf{x} = \mathbf{0}$ has only the trivial solution. Thus, $\ker T = \{(0, 0, 0)\}$. It follows from the dimension theorem that $\operatorname{im} T = \mathbb{R}^3$.

(e) $\begin{bmatrix} a & b \\ c & d \end{bmatrix} \in \ker T \Rightarrow a + b - c = 0, b + d = 0, a - c - d = 0$

$\Rightarrow c = a + b, d = -b.$

Hence, $\ker T = \left\{ \begin{bmatrix} a & b \\ a+b & -b \end{bmatrix} : a, b \in \mathbb{R} \right\}$

$= \left\{ a\begin{bmatrix} 1 & 0 \\ 1 & 0 \end{bmatrix} + b\begin{bmatrix} 0 & 1 \\ 1 & -1 \end{bmatrix} : a, b \in \mathbb{R} \right\}$

$= \left\langle \begin{bmatrix} 1 & 0 \\ 1 & 0 \end{bmatrix}, \begin{bmatrix} 0 & 1 \\ 1 & -1 \end{bmatrix} \right\rangle.$

Also, $\operatorname{im} T = \{(a+b-c, b+d, a-c-d): a, b, c, d \in \mathbb{R}\}$
$= \{a(1, 0, 1) + b(1, 1, 0) + c(-1, 0, -1)$
$\quad + d(0, 1, -1): a, b, c, d \in \mathbb{R}\}$
$= \langle (1, 0, 1), (1, 1, 0), (-1, 0, -1), (0, 1, -1) \rangle$
$= \langle (1, 0, 1), (1, 1, 0) \rangle.$

(It must be two-dimensional, by the dimension theorem; and these two vectors are linearly independent.)

(f) $\quad a_0 + a_1 x + a_2 x^2 \in \ker T \Leftrightarrow a_1 + 2a_2 x = 0 \Leftrightarrow a_1 = a_2 = 0.$

Hence, $\quad \ker T = \{a_0 : a_0 \in \mathbb{R}\}.$

Also, $\operatorname{im} T = \{a_1 + 2a_2 x : a_1, a_2 \in \mathbb{R}\} = \{b_1 + b_2 x : b_1, b_2 \in \mathbb{R}\}.$

2 Let $\mathbf{w} \in \operatorname{im} T$. Then there is a vector $\mathbf{v} \in V$ such that $\mathbf{w} = T(\mathbf{v})$. But $\mathbf{v} = \alpha_1 \mathbf{e}_1 + \ldots + \alpha_n \mathbf{e}_n$ for some $\alpha_1, \ldots, \alpha_n \in F$, and so
$$\mathbf{w} = T(\alpha_1 \mathbf{e}_1 + \ldots + \alpha_n \mathbf{e}_n) = \alpha_1 T(\mathbf{e}_1) + \ldots + \alpha_n T(\mathbf{e}_n).$$

3 The equivalence of (a) and (b) is clear; the equivalence of (c) and (d) follows from the dimension theorem; the equivalence of (b) and (c) follows from exercise 5.4.4.

4 (a) $(x, y) \in \ker T \Leftrightarrow 3x + 2y = x - 3y = 0 \Leftrightarrow x = y = 0.$ Therefore, nullity $T = 0$, rank $T = 2$, and T is bijective.
   (b) $(x, y, z) \in \ker T \Leftrightarrow x - y = x + z = 0 \Leftrightarrow y = x, z = -x.$ Therefore, $\ker T = \{(x, x, -x) : x \in \mathbb{R}\}$, nullity $T = 1$, rank $T = 2$, and T is surjective but not injective.
   (c) $(x, y) \in \ker T \Leftrightarrow x + y = 3x + 3y = 0 \Leftrightarrow y = -x.$ Therefore, $\ker T = \{(x, -x) : x \in \mathbb{R}\}$, nullity $T = 1$, rank $T = 1$, and T is neither surjective nor injective.
   (d) $(x, y) \in \ker T \Leftrightarrow (x, y, y) = (0, 0, 0) \Leftrightarrow x = y = 0.$ Therefore, $\ker T = \{(0, 0)\}$, nullity $T = 0$, rank $T = 2$, and T is injective but not surjective.

5 T is injective $\Leftrightarrow \operatorname{rank} T = \dim V$ (Theorem 6.4.5)
   $\qquad\qquad\qquad\qquad = \dim W$ (given)
   $\qquad\qquad \Leftrightarrow T$ is surjective. (Theorem 6.4.4).

6 Note that $S_\circ T : U \to W.$
   (a) Let $\mathbf{w} \in \operatorname{im}(S_\circ T)$. Then $\mathbf{w} = S_\circ T(\mathbf{u})$ for some $\mathbf{u} \in U$. Hence $\mathbf{w} = S(T(\mathbf{u})) \in \operatorname{im} S$, and so $\operatorname{im}(S_\circ T) \subset \operatorname{im} S$, from which the result follows.

(b) Define $R: \operatorname{im} T \to W$ by $R(\mathbf{v}) = (S_oT)(\mathbf{u})$ where $T(\mathbf{u}) = \mathbf{v}$. (This is the *restriction* of S to $\operatorname{im} T$, often denoted by $S|_{\operatorname{im} T}$.) Then $\operatorname{im} R = \operatorname{im}(S_oT)$, so rank $R = \operatorname{rank}(S_oT)$. Now R is injective, since S is injective, whence rank $R = \dim(\operatorname{im} T) = \operatorname{rank} T$.

(c) As in (b), rank $R = \operatorname{rank}(S_oT)$. But, if T is surjective, $\operatorname{im} T = V$, and so $R = S$.

Exercises 6.5

1 (a) The reduced echelon form is I_2, and so the rank is 2.
  (b) The reduced echelon form is I_3, and so the rank is 3.
  (c) The reduced echelon form is $\begin{bmatrix} 1 & 0 & 3 & 1 \\ 0 & 1 & -4 & -1 \\ 0 & 0 & 0 & 0 \\ 0 & 0 & 0 & 0 \end{bmatrix}$, and so the rank is 2.
  (d) The reduced echelon form is $\begin{bmatrix} 1 & 0 & 0 & -1 \\ 0 & 1 & 0 & 1 \\ 0 & 0 & 1 & 0 \end{bmatrix}$, and so the rank is 3.

2 (a) The required transformation $T: \mathbb{R}^2 \to \mathbb{R}^2$ is given by
$T((x, y)) = (x - y, 3x + 5y)$. Also
$\operatorname{im} T = \{(x - y, 3x + 5y): x, y \in \mathbb{R}\} = \{x(1, 3) + y(-1, 5): x, y \in \mathbb{R}\}$
$= \langle (1, 3), (-1, 5) \rangle$.

  (b) The required transformation $T: \mathbb{R}^3 \to \mathbb{R}^3$ is given by
$T((x, y, z)) = (-x - 2y + z, 2x + y + 2z, 5x + 2y + 2z)$.
Also, $\operatorname{im} T = \{x(-1, 2, 5) + y(-2, 1, 2) + z(1, 2, 2): x, y, z \in \mathbb{R}\}$
$= \langle (-1, 2, 5), (-2, 1, 2), (1, 2, 2) \rangle$.

  (c) The required transformation $T: \mathbb{R}^4 \to \mathbb{R}^4$ is given by
$T((x, y, z, w)) =$
$(4x + 3y + w, 3x + 2y + z + w, 7x + 5y + z + 2w, 6x + 5y - 2z + w)$.
Also $\operatorname{im} T = \langle (4, 3, 7, 6), (3, 2, 5, 5), (0, 1, 1, -2), (1, 1, 2, 1) \rangle$
$= \langle (4, 3, 7, 6), (3, 2, 5, 5) \rangle$.

  (d) The required transformation $T: \mathbb{R}^4 \to \mathbb{R}^3$ is given by
$T((x, y, z, w)) =$
$(3x + 5y - 2z + 2w, x + 2y - z + w, x + 2y - 2z + w)$.

Also, im $T = \langle(3,1,1),(5,2,2),(-2,-1,-2),(2,1,1)\rangle$
$= \langle(3,1,1),(5,2,2),(-2,-1,-2)\rangle$.

3 Let $\mathbf{A} = [a_{ij}]_{m \times n}$. Then

row space $A = \langle(a_{11},\ldots,a_{1n}),\ldots,(a_{i1},\ldots,a_{in}),\ldots,(a_{j1},\ldots,a_{jn}),\ldots,(a_{m1},\ldots,a_{mn})\rangle$
$= \langle(a_{11},\ldots,a_{1n}),\ldots,(a_{j1},\ldots,a_{jn}),\ldots,(a_{i1},\ldots,a_{in}),\ldots,(a_{m1},\ldots,a_{mn})\rangle$
$=$ row space $(E_{ij}A)$
$= \langle(a_{11},\ldots,a_{1n}),\ldots,\alpha(a_{i1},\ldots,a_{in}),\ldots,(a_{j1},\ldots,a_{jn}),$
$\ldots,(a_{m1},\ldots,a_{mn})\rangle (\alpha \neq 0)$
$=$ row space $(E_i(\alpha)A)$
$= \langle(a_{11},\ldots,a_{1n}),\ldots,(a_{i1}+\alpha a_{j1},\ldots,a_{in}+\alpha a_{jn}),$
$\ldots,(a_{j1},\ldots,a_{jn}),\ldots,(a_{m1},\ldots,a_{mn})\rangle$
$=$ row space $(E_{ij}(\alpha)A)$.

4 Simply change 'pre-multiplying' to 'post-multiplying', and 'e.r.o.' to 'e.c.o.'.

5 Let B be non-singular. Then $B = E_1 \ldots E_n$, where $E_1,\ldots,E_n$ are elementary matrices (see section 3.6), and $BA = E_1 \ldots E_n A$. The result now follows from Theorem 6.5.3(b) and a simple induction argument.

6 Let S, T be the linear transformations described at the beginning of section 6.5 and associated with A, B respectively. Then AB is associated with $S_0 T$, and

$$\text{rank}(AB) = \text{rank}(S_0 T) \leqslant \text{rank } S \quad \text{(exercise 6.4.6(a))}$$
$$= \text{rank } A.$$

7 There are the same number of non-zero rows in the echelon form as in the reduced echelon form.

8 This is straightforward.

Exercises 6.6

2 $\begin{vmatrix} 1 & 1 & 1 \\ a & a & 1 \\ a & 2 & 1 \end{vmatrix} = \begin{vmatrix} 1 & 0 & 0 \\ a & 0 & 1-a \\ a & 2-a & 1-a \end{vmatrix} = (a-1)(2-a) = 0 \Leftrightarrow a = 1, 2.$

If $a = 1$, the system has $x + y + z = 1$ and $x + y + z = 2$, and so is clearly inconsistent; if $a = 2$, the system has $2x + 2y + z = 3$ and $2x + 2y + z = 2$, and so is clearly inconsistent.

3
$$\begin{vmatrix} a & 2 & 1 \\ 2 & a & 1 \\ 1 & 1 & 1 \end{vmatrix} = \begin{vmatrix} a-1 & 1 & 0 \\ 1 & a-1 & 0 \\ 1 & 1 & 1 \end{vmatrix} = a(a-2) = 0 \Leftrightarrow a = 0, 2.$$

Hence there is a unique solution if $a \neq 0, 2$. Denote the matrix of coefficients of the system by A.

Then $A^{-1} = (1/a(a-2)) \begin{bmatrix} a-1 & -1 & 2-a \\ -1 & a-1 & 2-a \\ 2-a & 2-a & a^2-4 \end{bmatrix}$,

and $\begin{bmatrix} x \\ y \\ z \end{bmatrix} = A^{-1} \begin{bmatrix} 2a \\ -2 \\ 1 \end{bmatrix}$,

so the solution is

$$x = \frac{2a^2 - 3a + 4}{a(a-2)}, \quad y = \frac{4 - 5a}{a(a-2)}, \quad z = \frac{6a - a^2 - 8}{a(a-2)}.$$

If $a = 2$, the system has $2x + 2y + z = 4$ and $2x + 2y + z = -2$, which is clearly inconsistent.

Adding the first and second equations and putting $a = 0$ gives $2(x + y + z) = -2$, which is clearly inconsistent with $x + y + z = 1$, so the system is also inconsistent when $a = 0$.

4 Let $A = \begin{bmatrix} 3+a & 2+2a \\ 2a-3 & 2-a \\ 3 & 7 \end{bmatrix}$, $(A|B) = \begin{bmatrix} 3+a & 2+2a & 2-a \\ 2a-3 & 2-a & -3 \\ 3 & 7 & 1 \end{bmatrix}$.

Then rank $A = 2$, since the second column is not a multiple of the first. Hence, the system is consistent if and only if rank $(A|B) = 2$ (Theorem 6.6.1). The only other possible rank for $(A|B)$ is 3, so rank $(A|B) = 2$ if and only if $\det(A|B) = 0$. But $\det(A|B) = (22a + 1)(a - 3)$, so the system is consistent if and only if $a = 3, -\frac{1}{22}$. When $a = 3$, the solution is $x = -\frac{5}{6}, y = \frac{1}{2}$; when $a = -\frac{1}{22}$, the solution is $x = \frac{39}{47}, y = -\frac{10}{47}$.

7 EIGENVECTORS

7.1 CHANGING THE DOMAIN BASIS

Throughout this chapter V, W are vector spaces over the same field F. We have seen that the matrix associated with a given linear transformation $T: V \to W$ is dependent upon the bases chosen for V and W. Thus there are lots of such matrices: one corresponding to each choice of bases. It would be helpful for computational purposes if we knew how to make a particular choice of basis for which the associated matrix was as simple as possible. It is this objective that we have in mind in the current chapter.

We will try to describe the way in which the matrix associated with T changes when we change the basis in either, or both, of the domain and the codomain. Let us start with an example.

Example 7.1

Let $T: \mathbb{R}^3 \to \mathbb{R}^2$ be the linear transformation given by
$T((x, y, z)) = (x + 2y, y - z)$ and let A be the matrix associated with T with respect to the standard bases $(1, 0, 0), (0, 1, 0), (0, 0, 1)$ and $(1, 0), (0, 1)$ for $\mathbb{R}^3$ and $\mathbb{R}^2$ respectively. We will denote these bases briefly by a subscript d (for domain basis) and a subscript c (for codomain basis); we will also write $_cA_d$, rather than just A, when we wish to emphasise these bases. Then

$$T((1, 0, 0)) = (1, 0) = 1(1, 0) + 0(0, 1),$$
$$T((0, 1, 0)) = (2, 1) = 2(1, 0) + 1(0, 1),$$
$$T((0, 0, 1)) = (0, -1) = 0(1, 0) - 1(0, 1).$$

Thus $$_cA_d = \begin{bmatrix} 1 & 2 & 0 \\ 0 & 1 & -1 \end{bmatrix}.$$

Now, suppose that B represents T with respect to the bases $(1, 0, 0), (1, 1, 0),$

(1, 1, 1) and (1, 0), (0, 1). In similar manner denote these bases briefly by d' and c, and write $_cB_{d'}$ for B. Then

$$T((1,0,0)) = (1,0) = 1(1,0) + 0(0,1),$$
$$T((1,1,0)) = (3,1) = 3(1,0) + 1(0,1),$$
$$T((1,1,1)) = (3,0) = 3(1,0) + 0(0,1).$$

Therefore,
$$_cB_{d'} = \begin{bmatrix} 1 & 3 & 3 \\ 0 & 1 & 0 \end{bmatrix}.$$

What is the relationship between A and B? Let $J: \mathbb{R}^3 \to \mathbb{R}^3$ denote the identity map (see Theorem 6.2.3), and let $P(= {_dP_{d'}})$ be the matrix representing J with respect to the basis d' in domain and basis d in codomain. Then

$$J((1,0,0)) = (1,0,0) = 1(1,0,0) + 0(0,1,0) + 0(0,0,1),$$
$$J((1,1,0)) = (1,1,0) = 1(1,0,0) + 1(0,1,0) + 0(0,0,1),$$
$$J((1,1,1)) = (1,1,1) = 1(1,0,0) + 1(0,1,0) + 1(0,0,1).$$

Hence
$$_dP_{d'} = \begin{bmatrix} 1 & 1 & 1 \\ 0 & 1 & 1 \\ 0 & 0 & 1 \end{bmatrix}.$$

Now, $AP(= {_cA_d}\,{_dP_{d'}})$

$$= \begin{bmatrix} 1 & 2 & 0 \\ 0 & 1 & -1 \end{bmatrix} \begin{bmatrix} 1 & 1 & 1 \\ 0 & 1 & 1 \\ 0 & 0 & 1 \end{bmatrix} = \begin{bmatrix} 1 & 3 & 3 \\ 0 & 1 & 0 \end{bmatrix}$$

$$= B(= {_cB_{d'}}),$$

so it appears that we can find the matrix representing T with respect to bases d' in domain and c in codomain from that representing T with respect to bases d in domain and c in codomain by *post-multiplying* by the matrix P. In fact, this is true generally, as we will see next.

We call the matrix $P (= {_dP_{d'}})$ (namely the matrix representing the identity map with respect to bases d' in domain and d in codomain) the *transition matrix from* d' *to* d.

THEOREM 7.1.1 Let V, W be vector spaces of dimensions m, n respectively; let $e_1, \ldots, e_m$ and $e'_1, \ldots, e'_m$ be bases for V, let $f_1, \ldots, f_n$ be a basis for W, and denote these bases briefly by d, d' and c respectively; let $T: V \to W$ be a linear transformation. Suppose that $A (= {_cA_d})$, $B (= {_cB_{d'}})$ are matrices

representing T with respect to the bases indicated. Then
$$B = AP \qquad (_cB_{d'} = {}_cA_{dd}P_{d'}),$$
where P is the transition matrix from d' to d.

(*Note:* The suffices help us to remember to post-multiply by P, rather than to pre-multiply).

Proof This results by applying Theorem 6.3.1 in the following situation:

$$V \xrightarrow{J_V} V \xrightarrow{T} W$$
$$(d') \qquad (d) \qquad (c)$$

Since $T = T_o J_V$ we have (taking note of the relevant bases) $B = AP$.

EXERCISES 7.1

1 Let $T: \mathbb{R}^3 \to \mathbb{R}^3$ be the linear transformation given by
$$T((x, y, z)) = (x + y + z, 2x - y, 3y + 4z).$$
Find the matrix associated with T with respect to the standard basis in codomain and each of the following bases in domain:
 (a) standard basis;
 (b) $(1, 2, 3), (1, 1, 1), (0, 1, -1)$;
 (c) $(1, 1, 0), (0, 0, -1), (-1, 1, 0)$;
 (d) $(1, 0, 0), (1, 1, 0), (1, 1, 1)$.

2 In each of 1(b), (c), (d) above find the transition matrix from the given domain basis to the standard basis for $\mathbb{R}^3$. Use your result together with the matrix found in 1(a) to check your answers to 1(b), (c), (d).

3 The matrix $A = \begin{bmatrix} 1 & -1 \\ 3 & -5 \end{bmatrix}$ is associated with a linear transformation $T: P_1 \to P_1$ with respect to the basis $1, x$ in both domain and codomain. Write down the transition matrix P from the basis $1 + x, 1 - x$ to the basis $1, x$. Hence find the matrix associated with T with respect to the basis $1, x$ in codomain and $1 + x, 1 - x$ in domain.

4 Prove that every transition matrix is invertible.

7.2 CHANGING THE CODOMAIN BASIS

Now we shall consider how the matrix associated with T changes when we change the basis in the codomain. Let us consider the linear transformation T described in section 7.1 again.

Example 7.2

Let $T: \mathbb{R}^3 \to \mathbb{R}^2$ be given by $T((x, y, z)) = (x + 2y, y - z)$. Suppose that C represents T with respect to the bases $(1, 0, 0), (0, 1, 0), (0, 0, 1)$ for $\mathbb{R}^3$, and $(1, 0)$, $(1, 1)$ for $\mathbb{R}^2$ (and denote these briefly by d, c' respectively), and write $_{c'}C_d$. Then

$$T((1, 0, 0)) = (1, 0) = 1(1, 0) + 0(1, 1)$$
$$T((0, 1, 0)) = (2, 1) = 1(1, 0) + 1(1, 1)$$
$$T((0, 0, 1)) = (0, -1) = 1(1, 0) - 1(1, 1).$$

Therefore
$$_{c'}C_d = \begin{bmatrix} 1 & 1 & 1 \\ 0 & 1 & -1 \end{bmatrix}.$$

Again we seek the relationship between C and A. Let $Q (= {_cQ_{c'}})$ be the transition matrix from c' to c. Check that $Q = \begin{bmatrix} 1 & 1 \\ 0 & 1 \end{bmatrix}$. This time it is straightforward to check that $QC = A$. Since Q represents a linear transformation mapping a basis to a basis, it follows from Theorem 6.2.3 and Theorem 6.3.1 that Q is invertible. Hence

$$C = Q^{-1}A.$$

Once again this is a special case of a general result.

THEOREM 7.2.1 Adopt the same notation as in Theorem 7.1.1; let $\mathbf{f}'_1, \ldots, \mathbf{f}'_n$ be a basis for W and denote it by c'. Suppose that $A\ (= {_cA_d})$, $C\ (= {_{c'}C_d})$ are matrices representing T with respect to the bases indicated.
Then
$$C = Q^{-1}A \qquad ({_{c'}C_d} = {_{c'}Q^{-1}_c}\,{_cA_d}),$$
where $Q\ (= {_cQ_{c'}})$ is the transition matrix from c' to c.

Proof This time apply Theorem 6.3.1 in the following situation:

$$V \xrightarrow{T} W \xrightarrow{J_W} W$$
$$\text{(d)} \qquad \text{(c)} \qquad \text{(c')}$$

with T arrow also going from V to W (c').

Since $T = J_W \circ T$ (note the bases!), $C = Q^{-1}A$.

Note: We can find Q^{-1} from Q by the method of section 3.6, thus

$$\begin{bmatrix} 1 & 1 & | & 1 & 0 \\ 0 & 1 & | & 0 & 1 \end{bmatrix}$$
$$\downarrow$$
$$\begin{bmatrix} 1 & 0 & | & 1 & -1 \\ 0 & 1 & | & 0 & 1 \end{bmatrix} \quad R_1 = r_1 - r_2.$$

Therefore $\qquad Q^{-1} = \begin{bmatrix} 1 & -1 \\ 0 & 1 \end{bmatrix}.$

Alternatively, we can calculate Q^{-1} by considering the effect of J_W^{-1} ($= J_W$) on c, thus

$$J_W^{-1}((1,0)) = (1,0) = \quad 1(1,0) + 0(1,1)$$
$$J_W^{-1}((0,1)) = (0,1) = -1(1,0) + 1(1,1).$$

Hence $\qquad Q^{-1} = \begin{bmatrix} 1 & -1 \\ 0 & 1 \end{bmatrix}.$

EXERCISES 7.2

1 Let $T: \mathbb{R}^4 \to \mathbb{R}^2$ be the linear transformation given by

$$T((x, y, z, w)) = (x + y - w, 2y + 3z).$$

Find the matrix associated with T with respect to the standard basis in domain and each of the following bases in codomain:

(a) standard basis;
(b) $(1, 2), (0, -1)$;
(c) $(1, 1), (5, 4)$;
(d) $(-1, -3), (1, -5)$.

2 In each of 1(b), (c), (d) above find the transition matrix from the given codomain basis to the standard basis for $\mathbb{R}^2$. Use your result together with the matrix found in 1(a) to check your answers to 1(b), (c), (d).

3 The matrix $A = \begin{bmatrix} 1 & 1 & -1 \\ -2 & 3 & 1 \end{bmatrix}$ is associated with a linear transformation $T: P_2 \to \mathbb{C}$ with respect to the basis $1, x, x^2$ in domain and $1, i$ in codomain. Write down the transition matrix Q from the basis $1 + 2i, 3 - i$ to the basis $1, i$. Hence find the matrix associated with T with respect to the basis $1, x, x^2$ in domain and $1 + 2i, 3 - i$ in codomain.

7.3 CHANGING THE BASIS IN BOTH DOMAIN AND CODOMAIN

You can probably predict now what will happen to the matrix associated with T when we change the basis in both the domain and the codomain. We will state the general result.

THEOREM 7.3.1 Adopt the notation of Theorems 7.1.1, 7.2.1, and suppose that $_{c'}D_{d'}$ is the matrix representing T with respect to the bases indicated. Then
$$D = Q^{-1}AP \qquad (_{c'}D_{d'} = {}_{c'}Q^{-1}{}_{cc}A_{dd}P_{d'}).$$

Proof We will leave this as an exercise: the proofs of Theorems 7.1.1, 7.2.1 simply need to be amalgamated.

Let us check out the above result for the linear transformation T described in section 1.

Example 7.3

Let $T: \mathbb{R}^3 \to \mathbb{R}^2$ be given by $T((x, y, z)) = (x + 2y, y - z)$. If D represents T with respect to the bases $(1, 0, 0), (1, 1, 0), (1, 1, 1)$ for $\mathbb{R}^3$, and $(1, 0), (1, 1)$ for $\mathbb{R}^2$ (call them d', c' respectively), then
$$T((1, 0, 0)) = (1, 0) = 1(1, 0) + 0(1, 1)$$
$$T((1, 1, 0)) = (3, 1) = 2(1, 0) + 1(1, 1)$$
$$T((1, 1, 1)) = (3, 0) = 3(1, 0) + 0(1, 1).$$

Therefore $$_{c'}D_{d'} = \begin{bmatrix} 1 & 2 & 3 \\ 0 & 1 & 0 \end{bmatrix}.$$

You should check that this is $Q^{-1}AP$.

EXERCISES 7.3

Let $T: \mathbb{R}^3 \to \mathbb{R}^3$ be the linear transformation given by
$$T((x, y, z)) = (x + 2y - z, -z, 2x + y - 3z).$$

1 Write down the matrix associated with T with respect to the standard bases in domain and codomain.

2 By calculating the relevant transition matrices, find the matrices associated with T with respect to the following bases:

(a) $(-1, 0, 0)$, $(-1, -1, 0)$, $(-1, -1, -1)$ in both domain and codomain;
(b) $(1, 0, 0)$, $(1, 2, 0)$, $(1, 2, 3)$ in domain, $(1, 1, 1)$, $(1, 1, 0)$, $(1, 0, 0)$ in codomain;
(c) $(1, 1, 0)$, $(1, 0, 0)$, $(1, 1, 1)$ in domain, $(-1, 2, 1)$, $(0, 1, 2)$, $(0, 0, 1)$ in codomain.

3 Check your answers to 2 by calculating the matrices directly.

7.4 EIGENVALUES AND EIGENVECTORS

Consider now a linear transformation T from a vector space V into itself, and let A be a matrix associated with T with respect to the *same* basis in both domain and codomain. What is the simplest form that A can take, and how can we find a basis for V such that A takes this form? Well, A is certainly a square matrix, whichever basis we choose, and a particularly simple type of square matrix is a diagonal matrix, so it seems natural to ask when we can find a basis such that A is diagonal.

(Even better, of course, would be to make A a scalar matrix. However, this is too much to ask. For, suppose we can make a change of basis in V such that the matrix associated with T is scalar; let us suppose it becomes $B = \lambda I$. Then, since we are using the same basis in both domain and codomain, it follows from Theorem 7.3.1 that $A = P^{-1}BP = P^{-1}(\lambda I)P = \lambda P^{-1} P = \lambda I = B$. Hence, if A is not scalar to start with, no change of basis will make it scalar.)

In order to address this question, let us suppose that we can find a basis $\mathbf{e}_1, \ldots, \mathbf{e}_n$ for V such that the matrix associated with T is $A = \mathrm{diag}(a_{11}, \ldots, a_{nn})$.

Then
$$T(\mathbf{e}_i) = \sum_{j=1}^{n} a_{ji} \mathbf{e}_j = a_{ii} \mathbf{e}_i \quad \text{since } a_{ji} = 0 \text{ if } j \neq i.$$

We are thus interested in finding vectors $\mathbf{e}_i$ which are mapped to scalar multiples of themselves by T. Clearly the zero vector is such a vector, but it cannot be included in a basis for V, so is of no interest in this context. Consequently we define an *eigenvector* for T to be a *non-zero* vector $\mathbf{v} \in V$ such that $T(\mathbf{v}) = \lambda \mathbf{v}$ for some $\lambda \in F$. The field element λ is called an *eigenvalue* of T, and we speak of $\mathbf{v}$ as being an eigenvector *corresponding to* λ.

Examples 7.4

1. Consider the linear transformation $T: \mathbb{R}^2 \to \mathbb{R}^2$ given by $T((x, y)) = (x, -y)$. Geometrically this represents a reflection in the x-axis (Fig. 7.1).

   An eigenvector of T is a vector which lies along the same straight line as does its image under T. It is therefore clear that the eigenvectors are precisely those non-zero vectors lying along the x-axis or the y-axis.

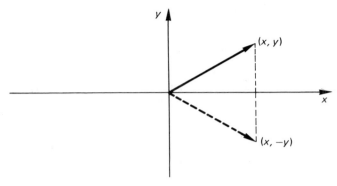

Fig. 7.1

Algebraically,

$$\lambda(x, y) = T((x, y)) = (x, -y) \Leftrightarrow \lambda x = x, \quad \lambda y = -y$$

$$\Leftrightarrow \lambda = 1 \quad \text{and} \quad y = 0,$$
$$\text{or } \lambda = -1 \quad \text{and} \quad x = 0.$$

Thus, the eigenvalues are 1 and -1, and the eigenvectors corresponding to $1, -1$ are

$$\{(x, 0): x \in \mathbb{R}, x \neq 0\}, \quad \{(0, y): y \in \mathbb{R}, y \neq 0\} \quad \text{respectively.}$$

2. Let $T: \mathbb{R}^2 \to \mathbb{R}^2$ be given by $T((x, y)) = (-y, x)$. Then T represents a rotation anticlockwise about the origin through an angle of $\frac{1}{2}\pi$ rad (Fig. 7.2). From this geometrical interpretation it is clear that no non-zero vector is mapped to a scalar multiple of itself. This linear transformation thus has no eigenvectors, and so the associated matrix is never diagonal, no matter which basis is chosen. The question of which linear transformations can be represented by a diagonal matrix therefore becomes an interesting one.

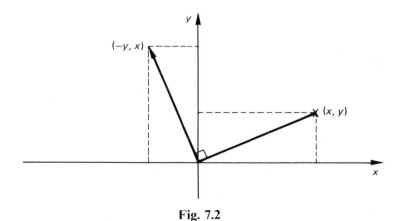

Fig. 7.2

3. Let λ be an eigenvalue for the linear transformation $T: V \to V$. Then there is an eigenvector $\mathbf{v} \in V$ such that $T(\mathbf{v}) = \lambda \mathbf{v}$. Hence
$$T^2(\mathbf{v}) = T(T(\mathbf{v})) = T(\lambda \mathbf{v}) = \lambda T(\mathbf{v}) = \lambda^2 \mathbf{v},$$
so $\mathbf{v}$ is also an eigenvector for T^2, and λ^2 is the corresponding eigenvalue. A straightforward induction argument shows that $\mathbf{v}$ is also an eigenvector for T^k ($k \geq 1$), and that λ^k is the corresponding eigenvalue.

If λ is an eigenvalue of T we denote by $E(\lambda)$ the set of eigenvectors corresponding to λ together with the zero vector; thus
$$E(\lambda) = \{\mathbf{v} \in V: T(\mathbf{v}) = \lambda \mathbf{v}\}.$$
In example 1 above,
$$E(1) = \{(x, 0): x \in \mathbb{R}\} = \text{the } x\text{-axis},$$
and
$$E(-1) = \{(0, y): y \in \mathbb{R}\} = \text{the } y\text{-axis}.$$
In each case, $E(\lambda)$ is a subspace of $\mathbb{R}^2$. A general result lies behind this fact.

THEOREM 7.4.1 Let λ be an eigenvalue for the linear transformation $T: V \to V$. Then $E(\lambda)$ is a subspace of V.

Proof The definition of λ implies that $E(\lambda)$ is non-empty. Suppose that $\mathbf{v}, \mathbf{w} \in E(\lambda)$. Then $T(\mathbf{v}) = \lambda \mathbf{v}$, $T(\mathbf{w}) = \lambda \mathbf{w}$,

and so $\qquad T(\mathbf{v} + \mathbf{w}) = T(\mathbf{v}) + T(\mathbf{w}) = \lambda \mathbf{v} + \lambda \mathbf{w} = \lambda(\mathbf{v} + \mathbf{w}),$

and $\qquad T(\alpha \mathbf{v}) = \alpha T(\mathbf{v}) = \alpha \lambda \mathbf{v} = \lambda(\alpha \mathbf{v}) \quad (\alpha \in F).$

Hence, $E(\lambda)$ is a subspace of V.

Consequently, we refer to $E(\lambda)$ as the *eigenspace* of λ. The eigenvectors corresponding to λ are the non-zero elements of the eigenspace of λ.

EXERCISES 7.4

1. Find the eigenvalues and corresponding eigenspaces for the following linear transformations:
   (a) $T: \mathbb{R}^2 \to \mathbb{R}^2$ given by $T((x, y)) = (-x, y)$;
   (b) $T: \mathbb{R}^2 \to \mathbb{R}^2$ given by $T((x, y)) = (y, x)$.

2. Show that the linear transformation $T: \mathbb{R}^2 \to \mathbb{R}^2$ given by
$$T((x, y)) = (x + y, -x + y)$$
has no real eigenvalues.

3 Find the eigenvalues of the linear transformation $T: \mathbb{R}^n \to \mathbb{R}^n$ which has associated matrix $[a_{ij}]_{n \times n}$, where $a_{ij} = 0$ for $i > j$, with respect to standard bases in both domain and codomain.

4 Let $T: V \to V$ be a linear transformation such that $T^r = J_V$, the identity map on V. Prove that if λ is an eigenvalue of T then $\lambda^r = 1$.

5 Suppose that $T: V \to V$ is a linear transformation such that $T^2 = J_V$, $T \neq J_V$, $T \neq -J_V$. By considering the vectors $\mathbf{v} + T(\mathbf{v})$ and $\mathbf{v} - T(\mathbf{v})$ for all $\mathbf{v} \in V$, show that $1, -1$ are eigenvalues of T, and show that these are the only eigenvalues of T.

7.5 THE CHARACTERISTIC EQUATION OF A SQUARE MATRIX

We saw in section 7.4 that linear transformations sometimes have eigenvectors and sometimes they do not. Here we seek a systematic method for determining whether or not T has eigenvalues and, if so, what they are, and what are the corresponding eigenspaces. Let us consider example 7.4.1 from the previous section again.

Example 7.5.1

$$T((x, y)) = \lambda(x, y) \Leftrightarrow (x, -y) = (\lambda x, \lambda y)$$
$$\Leftrightarrow x = \lambda x, \; -y = \lambda y$$
$$\Leftrightarrow \left.\begin{matrix} (\lambda - 1)x = 0 \\ (\lambda + 1)y = 0 \end{matrix}\right\} \quad (1)$$

So the problem of determining whether or not T has eigenvalues is equivalent to that of whether or not these equations have a non-trivial solution. As the equations are linear, we have already solved this problem in section 6.6. There is a non-trivial solution

$$\Leftrightarrow \begin{vmatrix} \lambda - 1 & 0 \\ 0 & \lambda + 1 \end{vmatrix} = 0$$
$$\Leftrightarrow (\lambda - 1)(\lambda + 1) = 0 \Leftrightarrow \lambda = \pm 1.$$

The corresponding eigenvectors can then be found by substituting these values for λ into (1) and solving these equations. Thus

$\lambda = 1$: $\left.\begin{matrix} 0x = 0 \\ 2y = 0 \end{matrix}\right\} \Leftrightarrow y = 0.$ Hence $E(1) = \{(x, 0): x \in \mathbb{R}\}.$

$\underline{\lambda = -1}$: $\left.\begin{array}{r}-2x = 0 \\ 0y = 0\end{array}\right\} \Leftrightarrow x = 0.$ Hence $E(-1) = \{(0, y): y \in \mathbb{R}\}$.

More generally, let T be any linear transformation from V to itself, and let A be a matrix associated with T with respect to some basis for V. Then

$$T(\mathbf{v}) = \lambda \mathbf{v} \Leftrightarrow A\mathbf{a}^T = \lambda \mathbf{a}^T \Leftrightarrow (A - \lambda I)\mathbf{a}^T = \mathbf{0},$$

where $\mathbf{a} = (\alpha_1, \ldots, \alpha_n)$, and $\alpha_1, \ldots, \alpha_n$ are the coordinates of $\mathbf{v}$ with respect to the basis for V. This system of equations has a non-trivial solution if and only if

$$\det(A - \lambda I) = 0.$$

The eigenvalues of T are, therefore, the values of λ for which $\det(A - \lambda I) = 0$. This matrix equation, which is a polynomial equation of degree n with coefficients in F, is called the *characteristic equation* of T (or of A), and the eigenvalues are, therefore, the roots in F of the characteristic equation.

We refer to the eigenvalues as being eigenvalues of A as well as of T. The eigenvectors of A corresponding to a given eigenvalue λ are the non-zero vectors $\mathbf{a} \in F^n$ (where A is $n \times n$) such that

$$A\mathbf{a}^T = \lambda \mathbf{a}^T,$$

or equivalently (taking the transpose of each side),

$$\mathbf{a}A^T = \lambda \mathbf{a}.$$

Note: It does not matter which matrix corresponding to T is chosen in calculating the characteristic equation. For, suppose that A, B both represent T with respect to different bases. Then $B = P^{-1}AP$ by Theorem 7.3.1, and $P^{-1}AP - \lambda I = P^{-1}(A - \lambda I)P$,

so $\qquad \det(B - \lambda I) = 0 \Leftrightarrow \det(P^{-1}(A - \lambda I)P) = 0$

$\qquad\qquad\qquad \Leftrightarrow \det(P^{-1})\det(A - \lambda I)\det P = 0$

$\qquad\qquad\qquad \Leftrightarrow \det(P^{-1})\det(P)\det(A - \lambda I) = 0$

$\qquad\qquad\qquad \Leftrightarrow \det(I)\det(A - \lambda I) = 0$

$\qquad\qquad\qquad \Leftrightarrow \det(A - \lambda I) = 0.$

Here then is the method we are seeking. Having found the eigenvalues, if any exist, the eigenspace corresponding to a particular eigenvalue λ can be found by solving the system of linear equations $A\mathbf{a}^T = \lambda \mathbf{a}^T$.

Examples 7.5.2

1. In example 7.5.1 the characteristic equation of T is $(\lambda - 1)(\lambda + 1) = 0$ (or $\lambda^2 - 1 = 0$).

2. Let $T: \mathbb{R}^2 \to \mathbb{R}^2$ be given by $T((x, y)) = (-y, x)$. Then with respect to standard bases in domain and codomain, T is represented by

$$A = \begin{bmatrix} 0 & -1 \\ 1 & 0 \end{bmatrix}.$$

Thus, $$0 = \det(A - \lambda I) = \begin{vmatrix} -\lambda & -1 \\ 1 & -\lambda \end{vmatrix} = \lambda^2 + 1$$

is the characteristic equation. Since this has no real solutions, T has no real eigenvalues.

3. Let $T: \mathbb{R}^2 \to \mathbb{R}^2$ be given by $T((x, y)) = (3x, 3y)$. Then with respect to standard bases in domain and codomain, T is represented by

$$A = \begin{bmatrix} 3 & 0 \\ 0 & 3 \end{bmatrix}.$$

Thus, $$0 = \det(A - \lambda I) = \begin{vmatrix} 3-\lambda & 0 \\ 0 & 3-\lambda \end{vmatrix} = (3-\lambda)^2$$

is the characteristic equation. This has the single real solution $\lambda = 3$, and so 3 is the only eigenvalue. To find the corresponding eigenspace we must solve the system $\begin{bmatrix} 3 & 0 \\ 0 & 3 \end{bmatrix} \begin{bmatrix} x \\ y \end{bmatrix} = 3 \begin{bmatrix} x \\ y \end{bmatrix}$; that is $\begin{matrix} 3x = 3x \\ 3y = 3y \end{matrix}$. These equations are clearly satisfied by all $x, y \in \mathbb{R}$, and so $E(3) = \mathbb{R}^2$.

EXERCISES 7.5

1 Find the eigenvalues, if any, and the corresponding eigenvectors for the following linear transformations:

(a) $T: \mathbb{R}^2 \to \mathbb{R}^2$ given by $T((x, y)) = (x + 3y, 2x + 2y)$;
(b) $T: P_1 \to P_1$ given by $T(a + bx) = (3a - b) + (a + b)x$;
(c) $T: \mathbb{C} \to \mathbb{C}$ given by $T(a + bi) = (a - 3b) + (a - b)i$.

2 Let $A = \begin{bmatrix} \cos\theta & -\sin\theta \\ \sin\theta & \cos\theta \end{bmatrix}$. Find the eigenvectors of A for the two cases

(a) θ is a multiple of π;
(b) θ is not a multiple of π.

3 Find the eigenvalues, if any, and corresponding eigenspaces for each of the following real matrices:

(a) $A = \begin{bmatrix} 2 & 1 & 1 \\ 1 & 2 & 1 \\ 1 & 1 & 2 \end{bmatrix}$; (b) $A = \begin{bmatrix} 2 & 0 & 0 \\ 0 & 1 & 2 \\ 0 & 0 & 2 \end{bmatrix}$;

(c) $A = \begin{bmatrix} 0 & 1 & 0 \\ 0 & 0 & 1 \\ -1 & -3 & -3 \end{bmatrix}$; (d) $A = \begin{bmatrix} 0 & 1 & 0 \\ 0 & 0 & 1 \\ 0 & -9 & 6 \end{bmatrix}$.

4 Let $P^{-1}AP$ be an $n \times n$ diagonal matrix with diagonal elements $a_{11}, \ldots, a_{nn}$. What are the eigenvalues of A? Justify your answer.

SOLUTIONS AND HINTS FOR EXERCISES

Exercises 7.1

1 (a) $\begin{bmatrix} 1 & 1 & 1 \\ 2 & -1 & 0 \\ 0 & 3 & 4 \end{bmatrix}$; (b) $\begin{bmatrix} 6 & 3 & 0 \\ 0 & 1 & -1 \\ 18 & 7 & -1 \end{bmatrix}$;

(c) $\begin{bmatrix} 2 & -1 & 0 \\ 1 & 0 & -3 \\ 3 & -4 & 3 \end{bmatrix}$; (d) $\begin{bmatrix} 1 & 2 & 3 \\ 2 & 1 & 1 \\ 0 & 3 & 7 \end{bmatrix}$.

2 (a) $\begin{bmatrix} 1 & 1 & 0 \\ 2 & 1 & 1 \\ 3 & 1 & -1 \end{bmatrix}$; (b) $\begin{bmatrix} 1 & 0 & -1 \\ 1 & 0 & 1 \\ 0 & -1 & 0 \end{bmatrix}$;

(c) $\begin{bmatrix} 1 & 1 & 1 \\ 0 & 1 & 1 \\ 0 & 0 & 1 \end{bmatrix}$.

3 $J(1+x) = 1 + x = 1.1 + 1.x$, $J(1-x) = 1 - x = 1.1 - 1.x$, so the transition matrix P is $\begin{bmatrix} 1 & 1 \\ 1 & -1 \end{bmatrix}$ and the matrix requested is $AP = \begin{bmatrix} 0 & 2 \\ -2 & 8 \end{bmatrix}$.

4 Every transition matrix is associated with respect to certain bases with the identity transformation, which is invertible. The result follows, therefore, from Theorem 6.3.1(d).

Exercises 7.2

1 (a) $\begin{bmatrix} 1 & 1 & 0 & -1 \\ 0 & 2 & 3 & 0 \end{bmatrix}$;

(b) $T((1,0,0,0)) = (1,0) = 1(1,2) + 2(0,-1)$;
$T((0,1,0,0)) = (1,2) = 1(1,2) + 0(0,-1)$;

$T((0, 0, 1, 0)) = (0, 3) = 0(1, 2) - 3(0, -1);$
$T((0, 0, 0, 1)) = (-1, 0) = -1(1, 2) - 2(0, -1).$ Matrix is $\begin{bmatrix} 1 & 1 & 0 & -1 \\ 2 & 0 & -3 & -2 \end{bmatrix}$.

(c) $(1, 0) = -4(1, 1) + 1(5, 4);$
$(1, 2) = 6(1, 1) - 1(5, 4);$ Matrix is $\begin{bmatrix} -4 & 6 & 15 & 4 \\ 1 & -1 & -3 & -1 \end{bmatrix}$.
$(0, 3) = 15(1, 1) - 3(5, 4);$
$(-1, 0) = 4(1, 1) - 1(5, 4).$

(d) $(1, 0) = -\frac{5}{8}(-1, -3) + \frac{3}{8}(1, -5);$
$(1, 2) = -\frac{7}{8}(-1, -3) + \frac{1}{8}(1, -5);$ Matrix is $\frac{1}{8}\begin{bmatrix} -5 & -7 & -3 & 5 \\ 3 & 1 & -3 & -3 \end{bmatrix}$.
$(0, 3) = -\frac{3}{8}(-1, -3) - \frac{3}{8}(1, -5);$
$(-1, 0) = \frac{5}{8}(-1, -3) - \frac{3}{8}(1, -5).$

2 (b) Transition matrix $Q = \begin{bmatrix} 1 & 0 \\ 2 & -1 \end{bmatrix}$; $Q^{-1} = \begin{bmatrix} 1 & 0 \\ 2 & -1 \end{bmatrix}$;

(c) $Q = \begin{bmatrix} 1 & 5 \\ 1 & 4 \end{bmatrix}$; $Q^{-1} = \begin{bmatrix} -4 & 5 \\ 1 & -1 \end{bmatrix}$;

(d) $Q = \begin{bmatrix} -1 & 1 \\ -3 & -5 \end{bmatrix}$; $Q^{-1} = \frac{1}{8}\begin{bmatrix} -5 & -1 \\ 3 & -1 \end{bmatrix}$.

3 The transition matrix $Q = \begin{bmatrix} 1 & 3 \\ 2 & -1 \end{bmatrix}$; $Q^{-1} = \frac{1}{7}\begin{bmatrix} 1 & 3 \\ 2 & -1 \end{bmatrix}$. Therefore, the new matrix is $Q^{-1}A = \frac{1}{7}\begin{bmatrix} -5 & 10 & 2 \\ 4 & -1 & -3 \end{bmatrix}$.

Exercises 7.3

1 The required matrix is $\begin{bmatrix} 1 & 2 & -1 \\ 0 & 0 & -1 \\ 2 & 1 & -3 \end{bmatrix}$.

2 (a) The transition matrix in both domain and codomain is

$P = \begin{bmatrix} -1 & -1 & -1 \\ 0 & -1 & -1 \\ 0 & 0 & -1 \end{bmatrix}$; $P^{-1} = \begin{bmatrix} -1 & 1 & 0 \\ 0 & -1 & 1 \\ 0 & 0 & -1 \end{bmatrix}$.

The required matrix is $P^{-1}AP = \begin{bmatrix} 1 & 3 & 3 \\ -2 & -3 & -1 \\ 2 & 3 & 0 \end{bmatrix}$.

(b) The transition matrices are

$$P = \begin{bmatrix} 1 & 1 & 1 \\ 0 & 2 & 2 \\ 0 & 0 & 3 \end{bmatrix}, \quad Q = \begin{bmatrix} 1 & 1 & 1 \\ 1 & 1 & 0 \\ 1 & 0 & 0 \end{bmatrix}; \quad Q^{-1} = \begin{bmatrix} 0 & 0 & 1 \\ 0 & 1 & -1 \\ 1 & -1 & 0 \end{bmatrix}.$$

The required matrix is $Q^{-1}AP = \begin{bmatrix} 2 & 4 & -5 \\ -2 & -4 & 2 \\ 1 & 5 & 5 \end{bmatrix}$.

(c) Here $P = \begin{bmatrix} 1 & 1 & 1 \\ 1 & 0 & 1 \\ 0 & 0 & 1 \end{bmatrix}, \quad Q = \begin{bmatrix} -1 & 0 & 0 \\ 2 & 1 & 0 \\ 1 & 2 & 1 \end{bmatrix}; \quad Q^{-1} = \begin{bmatrix} -1 & 0 & 0 \\ 2 & 1 & 0 \\ -3 & -2 & 1 \end{bmatrix}.$

The required matrix is $Q^{-1}AP = \begin{bmatrix} -3 & -1 & -2 \\ 6 & 2 & 3 \\ -6 & -1 & -4 \end{bmatrix}$.

3 (a) $T((-1,0,0)) = (-1,0,-2)$
$= 1(-1,0,0) - 2(-1,-1,0) + 2(-1,-1,-1),$
$T((-1,-1,0)) = (-3,0,-3)$
$= 3(-1,0,0) - 3(-1,-1,0) + 3(-1,-1,-1),$
$T((-1,-1,-1)) = (-2,1,0)$
$= 3(-1,0,0) - 1(-1,-1,0) + 0(-1,-1,-1).$

(b) $T((1,0,0)) = (1,0,2) = 2(1,1,1) - 2(1,1,0) + 1(1,0,0),$
$T((1,2,0)) = (5,0,4) = 4(1,1,1) - 4(1,1,0) + 5(1,0,0),$
$T((1,2,3)) = (2,-3,-5) = -5(1,1,1) + 2(1,1,0) + 5(1,0,0).$

(c) $T((1,1,0)) = (3,0,3) = -3(-1,2,1) + 6(0,1,2) - 6(0,0,1),$
$T((1,0,0)) = (1,0,2) = -1(-1,2,1) + 2(0,1,2) - 1(0,0,1),$
$T((1,1,1)) = (2,-1,0) = -2(-1,2,1) + 3(0,1,2) - 4(0,0,1).$

Exercises 7.4

1 (a) $\lambda(x,y) = T((x,y)) = (-x, y) \Leftrightarrow \lambda x = -x, \lambda y = y$
$\Leftrightarrow \lambda = 1$ and $x = 0$, or $\lambda = -1$ and $y = 0$.

Thus, the eigenvalues are 1 and -1, and the corresponding eigenvectors are

$\{(0, y): y \in \mathbb{R}, y \neq 0\}$ and $\{(x, 0): x \in \mathbb{R}, x \neq 0\}$ respectively.

(b) $\lambda(x, y) = T((x, y)) = (y, x) \Leftrightarrow \lambda x = y, \lambda y = x \Leftrightarrow (\lambda^2 - 1)x = 0, y = \lambda x$

$$\Leftrightarrow \lambda = \pm 1 \quad \text{and} \quad y = \pm x.$$

Thus, the eigenvalues are 1 and -1, and the corresponding eigenvectors are

$\{(x, x) : x \in \mathbb{R}, x \neq 0\}$ and $\{(x, -x) : x \in \mathbb{R}, x \neq 0\}$ respectively.

2 $\lambda(x, y) = T((x, y)) = (x + y, -x + y) \Leftrightarrow \lambda x = x + y, \lambda y = -x + y$

$$\Leftrightarrow (\lambda - 1)x - y = 0, x + (\lambda - 1)y = 0 \quad (1)$$

Now $\begin{vmatrix} \lambda - 1 & -1 \\ 1 & \lambda - 1 \end{vmatrix} = (\lambda - 1)^2 + 1 = \lambda^2 - 2\lambda + 2$, so the equations (1) have a non-trivial solution if and only if $\lambda^2 - 2\lambda + 2 = 0$. But this quadratic has no real roots, since $(-2)^2 - 4 \times 1 \times 2 = -4 < 0$.

3 $\lambda \mathbf{x} = T(\mathbf{x}) = \mathbf{x}A \Leftrightarrow \lambda x_1 = a_{11}x_1 + a_{12}x_2 + \ldots + a_{1n}x_n$

$\quad\quad\quad\quad\quad\quad\quad \lambda x_2 = \quad\quad a_{22}x_2 + \ldots + a_{2n}x_n$

$\quad\quad\quad\quad\quad\quad\quad \ldots\ldots\ldots\ldots\ldots\ldots\ldots\ldots\ldots$

$\quad\quad\quad\quad\quad\quad\quad \lambda x_n = \quad\quad\quad\quad\quad\quad a_{nn}x_n$

$\quad\quad\quad\Leftrightarrow (a_{11} - \lambda)x_1 + a_{12}x_2 + \ldots + a_{1n}x_n = 0$

$\quad\quad\quad\quad\quad (a_{22} - \lambda)x_2 + \ldots + a_{2n}x_n = 0$

$\quad\quad\quad\quad\quad \ldots\ldots\ldots\ldots\ldots\ldots\ldots\ldots$

$\quad\quad\quad\quad\quad\quad\quad\quad\quad\quad (a_{nn} - \lambda)x_n = 0.$

There is a non-trivial solution to this system

$$\Leftrightarrow \begin{vmatrix} a_{11} - \lambda & a_{12} & \ldots & a_{1n} \\ 0 & a_{22} - \lambda & \ldots & a_{2n} \\ \ldots & \ldots & \ldots & \ldots \\ 0 & 0 & \ldots & a_{nn} - \lambda \end{vmatrix} = 0$$

$\Leftrightarrow (a_{11} - \lambda)(a_{22} - \lambda)\ldots(a_{nn} - \lambda) = 0 \Leftrightarrow \lambda = a_{11}, a_{22}, \ldots,$ or a_{nn}.

4 λ is an eigenvalue $\Leftrightarrow T(\mathbf{x}) = \lambda \mathbf{x}$ and $\mathbf{x} \neq \mathbf{0}$

$\quad\quad\quad\quad\quad\quad\quad \Rightarrow \lambda^r \mathbf{x} = T^r(\mathbf{x}) = J_V(\mathbf{x}) = \mathbf{x}$, and $\mathbf{x} \neq \mathbf{0}$

$\quad\quad\quad\quad\quad\quad\quad \Rightarrow (\lambda^r - 1)\mathbf{x} = \mathbf{0}$ and $\mathbf{x} \neq \mathbf{0} \Rightarrow \lambda^r = 1.$

5 $T(\mathbf{v} + T(\mathbf{v})) = T(\mathbf{v}) + T^2(\mathbf{v}) = T(\mathbf{v}) + \mathbf{v}$, and 1 is an eigenvalue; $T(\mathbf{v} - T(\mathbf{v})) = T(\mathbf{v}) - T^2(\mathbf{v}) = T(\mathbf{v}) - \mathbf{v} = -(\mathbf{v} - T(\mathbf{v}))$, and -1 is an eigenvalue.

Also, λ is an eigenvalue $\Rightarrow \lambda^2 = 1$ (exercise 7.4.4) $\Rightarrow \lambda = \pm 1$.

Exercises 7.5

1. (a) With respect to standard bases in domain and codomain, T is represented by $A = \begin{bmatrix} 1 & 3 \\ 2 & 2 \end{bmatrix}$. The eigenvalues are $4, -1$ and the corresponding eigenspaces are $\{(x, x): x \in \mathbb{R}\}$ and $\{(3x, -2x): x \in \mathbb{R}\}$ respectively.

    (b) With respect to basis $1, x$ in both domain and codomain, T is represented by $A = \begin{bmatrix} 3 & -1 \\ 1 & 1 \end{bmatrix}$. The eigenvalue is 2 and the corresponding eigenspace is $\{a + ax: a \in \mathbb{R}\}$.

    (c) With respect to the basis $1, i$ in both domain and codomain, T is represented by
    $$A = \begin{bmatrix} 1 & -3 \\ 1 & -1 \end{bmatrix}.$$
    This has no real eigenvalues.

2. The characteristic equation of A is $\lambda^2 - 2\lambda \cos\theta + 1 = 0$, and the eigenvalues are $\lambda = \cos\theta \pm i\sin\theta$.

    (a) If $\theta = n\pi$, then $\sin\theta = 0$ and $\lambda = \cos\theta$. The eigenspace corresponding to this eigenvalue is $\mathbb{R}^2$.

    (b) If $\theta \neq n\pi$, then the eigenspace corresponding to λ is $\{k(\pm i, 1): k \in \mathbb{C}\}$. Hence, there are no real eigenvectors.

3. (a) The eigenvalues are $1, 4$ and the corresponding eigenspaces are
    $$E(1) = \{(a + b, -a, -b): a, b \in \mathbb{R}\}, \qquad E(4) = \{(a, a, a): a \in \mathbb{R}\}.$$

    (b) The eigenvalues are $1, 2$ and the corresponding eigenspaces are
    $$E(1) = \{(0, a, 0): a \in \mathbb{R}\}, \qquad E(2) = \{(a, 2b, b): a, b \in \mathbb{R}\}.$$

    (c) The only eigenvalue is -1 and the corresponding eigenspace is $E(-1) = \{(a, -a, a): a \in \mathbb{R}\}$.

    (d) The eigenvalues are $0, 3$ and the corresponding eigenspaces are
    $$E(0) = \{(a, 0, 0): a \in \mathbb{R}\}, \qquad E(3) = \{(a, 3a, 9a): a \in \mathbb{R}\}.$$

4. The eigenvalues are $a_{11}, \ldots, a_{nn}$. For, the eigenvalues of A are the same as those of $P^{-1}AP$ (see the note in the text of section 7.5).

8 ORTHOGONAL REDUCTION OF SYMMETRIC MATRICES

8.1 ORTHOGONAL VECTORS AND MATRICES

In this final chapter we will give an application of some of the techniques introduced in this book to the study of equations of degree two. We started the book by considering the geometry of lines and planes; we will see now how some of the more powerful ideas that have resulted by introducing greater abstraction can be employed to solve more complex problems in geometry. Thus, in a sense, we will have come full circle.

First we need to generalise the important concepts in section 1.3. The *scalar product*, $\mathbf{x} \cdot \mathbf{y}$, or $\mathbf{x}, \mathbf{y} \in \mathbb{R}^n$ is defined to be the matrix product $\mathbf{x}\mathbf{y}^T$. Thus, if $\mathbf{x} = (x_1, \ldots, x_n)$, $\mathbf{y} = (y_1, \ldots, y_n)$,

$$\mathbf{x} \cdot \mathbf{y} = (x_1, \ldots, x_n) \begin{bmatrix} y_1 \\ \vdots \\ y_n \end{bmatrix} = x_1 y_1 + \ldots + x_n y_n = \sum_{i=1}^{n} x_i y_i.$$

Notice that $\mathbf{x} \cdot \mathbf{y} = \mathbf{y} \cdot \mathbf{x}$, and that $\mathbf{x} \cdot \mathbf{x} = \sum_{i=1}^{n} x_i^2$.

The *length* of $\mathbf{x}$ is defined to be the non-negative real number $\sqrt{\mathbf{x} \cdot \mathbf{x}} = \sqrt{\sum_{i=1}^{n} x_i^2}$; we denote it by $|\mathbf{x}|$. A vector of length 1 is called a *unit* vector.

Also, $\dfrac{\mathbf{x} \cdot \mathbf{y}}{|\mathbf{x}||\mathbf{y}|}$ is called the *cosine of the angle* between $\mathbf{x}$ and $\mathbf{y}$.

If this cosine is zero (that is, if $\mathbf{x} \cdot \mathbf{y} = 0$) then $\mathbf{x}$ and $\mathbf{y}$ are said to be *orthogonal*. Finally we define an $n \times n$ real matrix A to be orthogonal if $A^T A = I_n$.

Examples 8.1

1. Let $\mathbf{x} = (1, -1, 0, 2, 5)$, $\mathbf{y} = (0, 0, 4, -1, -1)$. Then

   $\mathbf{x} \cdot \mathbf{y} = 1 \times 0 + (-1) \times 0 + 0 \times 4 + 2 \times (-1) + 5 \times (-1) = -7,$

$|\mathbf{x}| = \sqrt{31}$, $|\mathbf{y}| = \sqrt{18}$,

$\cos\theta = (-7)/(\sqrt{31}\sqrt{18})$, where θ is the angle between $\mathbf{x}$ and $\mathbf{y}$.

2. Let $A = \begin{bmatrix} 1/\sqrt{2} & -1/\sqrt{2} \\ 1/\sqrt{2} & 1/\sqrt{2} \end{bmatrix}$. Then $A^T A = \begin{bmatrix} 1 & 0 \\ 0 & 1 \end{bmatrix}$, and so A is orthogonal.

3. Let A, B be orthogonal $n \times n$ matrices. Then AB is an $n \times n$ matrix and $(AB)^T(AB) = B^T A^T AB = B^T I_n B = B^T B = I_n$, so AB is also orthogonal.

LEMMA 8.1.1 If A is an $n \times n$ orthogonal matrix, then
(a) $\det A = \pm 1$; (b) $A^{-1} = A^T$.

Proof

(a) We have $1 = \det I_n = \det(A^T A) = (\det A^T)(\det A) = (\det A)^2$,

which implies that $\det A = \pm 1$.

(b) Part (a) shows that A is non-singular, so A^{-1} exists. Moreover,
$$A^{-1} = I_n A^{-1} = A^T A A^{-1} = A^T.$$

COROLLARY 8.1.2 A is orthogonal if and only if $AA^T = I_n$.

Proof A is orthogonal $\Leftrightarrow A^T A = I_n \Leftrightarrow (A^T)^{-1} A^T A A^{-1} = (A^T)^{-1} A^{-1}$
$$\Leftrightarrow I_n = (A^{-1})^T A^{-1} = (A^T)^T A^T = AA^T.$$

COROLLARY 8.1.3 A is orthogonal if and only if A^T is orthogonal.

Proof A is orthogonal $\Leftrightarrow A^T A = I_n \Leftrightarrow A^T(A^T)^T = I_n$

$\Leftrightarrow A^T$ is orthogonal (by Corollary 8.1.2).

Using the above results we can derive the following characterisation of orthogonal matrices; it is interesting because of the geometric information it gives us concerning the linear transformations associated with such matrices.

THEOREM 8.1.4 If A is an $n \times n$ matrix, the following are equivalent:
(a) A is orthogonal;

(b) A preserves the lengths of vectors (i.e., $|\mathbf{v}A| = |\mathbf{v}|$ for all $\mathbf{v} \in \mathbb{R}^n$);
(c) A preserves scalar products (i.e., $\mathbf{v} \cdot \mathbf{w} = (\mathbf{v}A) \cdot (\mathbf{w}A)$ for all $\mathbf{v}, \mathbf{w} \in \mathbb{R}^n$).

Proof

(a)$\Rightarrow$(b) Let $\mathbf{v} \in \mathbb{R}^n$. Then
$$|\mathbf{v}A|^2 = (\mathbf{v}A) \cdot (\mathbf{v}A) = (\mathbf{v}A)(\mathbf{v}A)^T = \mathbf{v}AA^T\mathbf{v}^T = \mathbf{v}I_n\mathbf{v}^T = \mathbf{v}\mathbf{v}^T = |\mathbf{v}|^2.$$
Since $|\mathbf{v}A| \geq 0$, $|\mathbf{v}| \geq 0$, we have $|\mathbf{v}A| = |\mathbf{v}|$.

(b)$\Rightarrow$(c) Let $\mathbf{v}, \mathbf{w} \in \mathbb{R}^n$. First we show that
$$|\mathbf{v} - \mathbf{w}|^2 = |\mathbf{v}|^2 + |\mathbf{w}|^2 - 2\mathbf{v} \cdot \mathbf{w}. \tag{1}$$

For
$$\begin{aligned}|\mathbf{v} - \mathbf{w}|^2 &= (\mathbf{v} - \mathbf{w})(\mathbf{v} - \mathbf{w})^T \\ &= (\mathbf{v} - \mathbf{w})(\mathbf{v}^T - \mathbf{w}^T) \\ &= \mathbf{v}\mathbf{v}^T - \mathbf{v}\mathbf{w}^T - \mathbf{w}\mathbf{v}^T + \mathbf{w}\mathbf{w}^T \\ &= |\mathbf{v}|^2 - \mathbf{v} \cdot \mathbf{w} - \mathbf{w} \cdot \mathbf{v} + |\mathbf{w}|^2 \\ &= |\mathbf{v}|^2 + |\mathbf{w}|^2 - 2\mathbf{v} \cdot \mathbf{w}, \text{ simply using properties of matrices.}\end{aligned}$$

Using (1) then we have
$$|\mathbf{v} - \mathbf{w}|^2 = |\mathbf{v}|^2 + |\mathbf{w}|^2 - 2\mathbf{v} \cdot \mathbf{w}$$
and
$$|(\mathbf{v} - \mathbf{w})A|^2 = |\mathbf{v}A - \mathbf{w}A|^2 = |\mathbf{v}A|^2 + |\mathbf{w}A|^2 - 2(\mathbf{v}A) \cdot (\mathbf{w}A).$$
Since we are assuming that (b) holds,
$$|\mathbf{v}|^2 = |\mathbf{v}A|^2, \qquad |\mathbf{w}|^2 = |\mathbf{w}A|^2, \qquad |\mathbf{v} - \mathbf{w}|^2 = |(\mathbf{v} - \mathbf{w})A|^2,$$
and so $2\mathbf{v} \cdot \mathbf{w} = 2(\mathbf{v}A) \cdot (\mathbf{w}A)$. Hence (c) holds.

(c)$\Rightarrow$(d) Let $\mathbf{e}_1, \ldots, \mathbf{e}_n$ be the standard basis for $\mathbb{R}^n$. Write $B = AA^T$. Then, since (c) holds,
$$\mathbf{e}_i \mathbf{e}_j^T = (\mathbf{e}_i A)(\mathbf{e}_j A)^T = \mathbf{e}_i AA^T \mathbf{e}_j^T = \mathbf{e}_i B \mathbf{e}_j^T.$$
Writing this out fully, we have

$$\mathbf{e}_i \mathbf{e}_j^T = [0 \ \ldots \ 0 \ 1 \ 0 \ \ldots \ 0] \begin{bmatrix} b_{11} & \ldots & b_{1n} \\ \ldots & & \ldots \\ b_{n1} & \ldots & b_{nn} \end{bmatrix} \begin{bmatrix} 0 \\ \vdots \\ 0 \\ 1 \\ 0 \\ \vdots \\ 0 \end{bmatrix} \leftarrow j\text{th position}$$

$\uparrow$
ith position

$$= [b_{i1} \quad \cdots \quad b_{in}] \begin{bmatrix} 0 \\ \vdots \\ 0 \\ 1 \\ 0 \\ \vdots \\ 0 \end{bmatrix} \leftarrow j\text{th position}$$

$$= [b_{ij}].$$

But $\mathbf{e}_i \mathbf{e}_j^T = \begin{cases} 1 & \text{if } i = j \\ 0 & \text{if } i \neq j \end{cases}$. Thus, $b_{ij} = \begin{cases} 1 & \text{if } i = j \\ 0 & \text{if } i \neq j \end{cases}$; that is, $AA^T = B = I_n$.

EXERCISES 8.1

1 Let θ be the angle between the vectors $\mathbf{v}_1 = (1, 2, 3, 4)$, $\mathbf{v}_2 = (0, -1, -1, 2)$. Find $\cos \theta$. Let $\mathbf{v}_3 = (0, 0, 0, 1)$. Find a vector $\mathbf{v}$ which is orthogonal to $\mathbf{v}_1, \mathbf{v}_2, \mathbf{v}_3$.

2 Let $\mathbf{v}_1, \mathbf{v}_2, \mathbf{v}_3, \mathbf{v}_4$ be mutually orthogonal unit vectors in $\mathbb{R}^4$ (that is, $\mathbf{v}_i \cdot \mathbf{v}_i = 1$, $\mathbf{v}_i \cdot \mathbf{v}_j = 0$ when $i \neq j$). Prove that they are linearly independent.

3 Let A, B be $n \times n$ orthogonal matrices. Prove that AB^T is an $n \times n$ orthogonal matrix.

4 The $n \times n$ matrix A has the property that, for every $\mathbf{u}, \mathbf{v} \in \mathbb{R}^n$, $(\mathbf{u}A)(\mathbf{v}A)^T = \mathbf{u}\mathbf{v}^T$. Show that A is orthogonal.

5 The $n \times n$ matrix A is said to be *congruent* to the $n \times n$ matrix B if there is an $n \times n$ orthogonal matrix P such that $P^T B P = A$. Prove that

(a) every $n \times n$ matrix is congruent to itself;
(b) if A is congruent to B, then B is congruent to A;
(c) if A, B, C are $n \times n$ matrices such that A is congruent to B, and B is congruent to C, then A is congruent to C.

8.2 EUCLIDEAN TRANSFORMATIONS

A transformation $T: \mathbb{R}^n \to \mathbb{R}^n$ is called *Euclidean* if it preserves distances between vectors in $\mathbb{R}^n$; that is, if $|T(\mathbf{v}) - T(\mathbf{w})| = |\mathbf{v} - \mathbf{w}|$ for all $\mathbf{v}, \mathbf{w} \in \mathbb{R}^n$.

Examples 8.2

1. Let a, b be fixed real numbers. A transformation $T: \mathbb{R}^2 \to \mathbb{R}^2$ given by

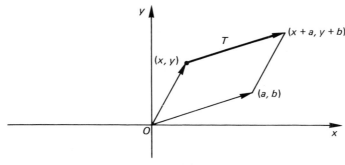

Fig. 8.1

$T((x, y)) = (x + a, y + b)$ is called a *translation*. Geometrically, the effect is as shown in Fig. 8.1.

Let $\mathbf{v} = (x_1, y_1)$, $\mathbf{w} = (x_2, y_2)$. Then $\mathbf{v} - \mathbf{w} = (x_1 - x_2, y_1 - y_2)$,

$T(\mathbf{v}) - T(\mathbf{w}) = ((x_1 + a) - (x_2 + a), (y_1 + b) - (y_2 + b)) = (x_1 - x_2, y_1 - y_2)$,

and so T is clearly euclidean.

2. Consider the transformation $R_\theta \colon \mathbb{R}^2 \to \mathbb{R}^2$ obtained by rotating all vectors in the plane anticlockwise through an angle θ about the origin (Fig. 8.2).

Then
$$\begin{aligned}x' &= OP' \cos(\theta + \phi) \\ &= OP'(\cos\theta \cos\phi - \sin\theta \sin\phi) \\ &= (OP \cos\phi)\cos\theta - (OP \sin\phi)\sin\theta \\ &= x\cos\theta - y\sin\theta.\end{aligned}$$

Similarly,
$$\begin{aligned}y' &= OP' \sin(\theta + \phi) \\ &= OP'(\sin\theta \cos\phi + \cos\theta \sin\phi) \\ &= (OP \cos\phi)\sin\theta + (OP \sin\phi)\cos\theta \\ &= x\sin\theta + y\cos\theta.\end{aligned}$$

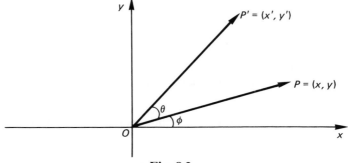

Fig. 8.2

Thus, R_θ is given by $R_\theta((x, y)) = (x \cos \theta - y \sin \theta, x \sin \theta + y \cos \theta)$.

A straightforward calculation shows that R_θ is a Euclidean transformation.

3. Let $E: \mathbb{R}^2 \to \mathbb{R}^2$ be the linear transformation given by $E((x, y)) = (x, -y)$ (see example 7.4.1). Then E represents a reflection in the x-axis, and is easily seen to be a euclidean transformation.

Euclidean transformations of the plane are important because they preserve the geometric properties of planar figures: for example, they map circles to circles, and triangles to congruent triangles. It can be shown that such a plane transformation is Euclidean if and only if it is

(a) a Euclidean linear transformation,
(b) a translation, or
(c) a composite of any two transformations from (a) or (b).

We will not prove this result, but will complete the classification of Euclidean transformations of the plane by determining the linear ones. First we relate Euclidean linear transformations to orthogonal matrices.

THEOREM 8.2.1 A linear transformation $T: \mathbb{R}^n \to \mathbb{R}^n$ is Euclidean if and only if the matrix associated with it with respect to the standard basis vectors is orthogonal.

Proof Let $T: \mathbb{R}^n \to \mathbb{R}^n$ be a linear transformation, let A be the matrix associated with T with respect to the standard basis for $\mathbb{R}^n$, and let $\mathbf{v} = (v_1, \ldots, v_n), \mathbf{w} = (w_1, \ldots, w_n) \in \mathbb{R}^n$. Then $T(\mathbf{v}) = \mathbf{v} A^T$, $T(\mathbf{w}) = \mathbf{w} A^T$ (see section 6.3). Hence

T is Euclidean $\Leftrightarrow |\mathbf{v} - \mathbf{w}| = |T(\mathbf{v}) - T(\mathbf{w})| = |T(\mathbf{v} - \mathbf{w})|$ for all $\mathbf{v}, \mathbf{w} \in \mathbb{R}^n$

$\Leftrightarrow |\mathbf{v} - \mathbf{w}| = |(\mathbf{v} - \mathbf{w}) A^T|$ for all $\mathbf{v}, \mathbf{w} \in \mathbb{R}^n$

$\Leftrightarrow |\mathbf{x}| = |\mathbf{x} A^T|$ for all $\mathbf{x} \in \mathbb{R}^n$

$\Leftrightarrow A^T$ is orthogonal   (by Theorem 8.1.4)

$\Leftrightarrow A$ is orthogonal   (by Corollary 8.1.3).

We have thus reduced the problem of classifying all euclidean transformations to that of finding all orthogonal matrices. This can be solved quite easily.

THEOREM 8.2.2 Let A be a 2×2 orthogonal matrix. Then A has the form

$$\begin{bmatrix} \cos \theta & -\sin \theta \\ \sin \theta & \cos \theta \end{bmatrix} \quad \text{or} \quad \begin{bmatrix} \cos \theta & \sin \theta \\ \sin \theta & -\cos \theta \end{bmatrix}.$$

Proof It is easily checked that the two types of matrix given are orthogonal. So suppose that $A = \begin{bmatrix} a & b \\ c & d \end{bmatrix}$ is orthogonal. Then

$$\begin{bmatrix} 1 & 0 \\ 0 & 1 \end{bmatrix} = A^T A = \begin{bmatrix} a & c \\ b & d \end{bmatrix}\begin{bmatrix} a & b \\ c & d \end{bmatrix} = \begin{bmatrix} a^2+c^2 & ab+cd \\ ba+dc & b^2+d^2 \end{bmatrix}.$$

Hence A is orthogonal if and only if

$$a^2 + c^2 = b^2 + d^2 = 1 \qquad (1)$$

and
$$ab + cd = 0. \qquad (2)$$

Now equations (1) imply that there is a $\theta \in \mathbb{R}$ such that $a = \cos\theta, c = \sin\theta$. But then (2) implies that $b\cos\theta + d\sin\theta = 0$; that is,

$$\tan\theta = -b/d. \qquad (3)$$

Substituting $-d\tan\theta$ for b in (1) we get

$$1 = d^2\tan^2\theta + d^2 = d^2(1+\tan^2\theta) = d^2\sec^2\theta.$$

Thus $d^2 = \cos^2\theta$; that is, $d = \pm\cos\theta$. From (3) it follows that $b = -\sin\theta$ when $d = \cos\theta$, and $b = \sin\theta$ when $d = -\cos\theta$. Therefore A has one of the two stated forms.

COROLLARY 8.2.3 Let $T: \mathbb{R}^2 \to \mathbb{R}^2$ be a Euclidean linear transformation of the plane. Then either $T = R_\theta$ or $T = ER_\theta$ for some angle θ. (Here the notation is as in examples 2 and 3 above.) Furthermore, if A is the matrix associated with T with respect to the standard basis vectors for $\mathbb{R}^2$, then $T = R_\theta$ if and only if $\det A = 1$, and $T = ER_\theta$ if and only if $\det A = -1$.

Proof The matrix A is orthogonal (by Theorem 8.2.1). Hence, by Theorem 8.2.2,

$$A = \begin{bmatrix} \cos\theta & -\sin\theta \\ \sin\theta & \cos\theta \end{bmatrix} \quad \text{or} \quad A = \begin{bmatrix} \cos\theta & \sin\theta \\ \sin\theta & -\cos\theta \end{bmatrix}.$$

If the former holds, then $\det A = 1$ and

$$T((x,y)) = (x,y)A^T = (x \;\; y)\begin{bmatrix} \cos\theta & \sin\theta \\ -\sin\theta & \cos\theta \end{bmatrix} = (x\cos\theta - y\sin\theta, x\sin\theta + y\cos\theta)$$

$$= R_\theta((x,y)).$$

If the latter holds, then $\det A = -1$ and

$$T((x,y)) = (x,y)A^T = (x \;\; y)\begin{bmatrix} \cos\theta & \sin\theta \\ \sin\theta & -\cos\theta \end{bmatrix} = (x\cos\theta + y\sin\theta, x\sin\theta - y\cos\theta)$$

$$= ER_{-\theta}((x,y)) = ER_{\theta'}((x,y)), \qquad \text{where } \theta' = 2\pi - \theta.$$

In a similar manner Euclidean transformations of three-dimensional space can be classified. Again they are linear, translations, or composites of two such. Moreover, we have the following analogue of Corollary 8.2.3.

THEOREM 8.2.4 Let $T: \mathbb{R}^3 \to \mathbb{R}^3$ be a Euclidean transformation of space, and let A be the matrix associated with T with respect to the standard basis vectors for $\mathbb{R}^3$. Then either

(a) $\det A = 1$ and T represents a rotation about some axis through the origin, or
(b) $\det A = -1$ and T represents a rotation followed by a reflection in a plane through the origin.

EXERCISES 8.2

1 Show that E (see example 8.2.3) is a Euclidean linear transformation.

2 Show that R_θ (see example 8.2.2) is a Euclidean linear transformation.

3 Let $r_\theta: \mathbb{R}^3 \to \mathbb{R}^3$ be the transformation obtained by rotating all vectors in $\mathbb{R}^3$ anticlockwise through an angle θ about the z-axis. Find the matrix associated with r_θ with respect to the standard basis for $\mathbb{R}^3$, and check that it is orthogonal.

4 A basis $\mathbf{e}_1, \ldots, \mathbf{e}_n$ for $\mathbb{R}^n$ is *orthonormal* if $\mathbf{e}_i \cdot \mathbf{e}_i = 1$ ($1 \leq i \leq n$) and $\mathbf{e}_i \cdot \mathbf{e}_j = 0$ ($i \neq j$). Prove that the linear transformation $T: \mathbb{R}^n \to \mathbb{R}^n$ is Euclidean if and only if T maps an orthonormal basis for $\mathbb{R}^n$ to an orthonormal basis for $\mathbb{R}^n$.

5 Prove that a Euclidean linear transformation is associated with an orthogonal matrix with respect to any orthonormal basis for $\mathbb{R}^n$.

8.3 ORTHOGONAL REDUCTION OF A REAL SYMMETRIC MATRIX

Recall from section 2.5 that a square matrix A is symmetric if $A = A^T$. We saw in section 7.4 that not every linear transformation $T: V \to V$ can be represented by a diagonal matrix. However, if $T: \mathbb{R}^n \to \mathbb{R}^n$ is a linear transformation whose associated matrix with respect to the standard basis for $\mathbb{R}^n$ is symmetric, then T can be so represented. We will prove this result here for the cases $n = 2, 3$, as it has important consequences for the geometry of the plane and of space.

In matrix terms what we are claiming is that if A is symmetric then there is a non-singular matrix P such that $P^{-1}AP$ is diagonal (where P is a suitable transition matrix). In fact, it is possible to choose P to be orthogonal; that is, if

A is symmetric then there is an orthogonal matrix P such that $P^T A P$ is diagonal (since $P^{-1} = P^T$ when P is orthogonal).

First we need two results concerning the eigenvectors of a real symmetric matrix. Note that $\boldsymbol{a}$ is an eigenvector corresponding to λ if and only if $\boldsymbol{a} \neq \boldsymbol{0}$ and $A\boldsymbol{a}^T = \lambda \boldsymbol{a}^T$ (or equivalently $\boldsymbol{a}A = \lambda \boldsymbol{a}$, since $A = A^T$)—see section 7.5.

THEOREM 8.3.1   The eigenvalues of a real symmetric matrix are all real.

Proof   Let A be a real symmetric $n \times n$ matrix, let λ be an eigenvalue of A, and let $\mathbf{v} = (v_1, \ldots, v_n)$ be an eigenvector of A corresponding to λ. Then $\mathbf{v}A = \lambda \mathbf{v}$. Now if B is any matrix, denote by $\bar{B}$ the matrix whose (i,j)-element is the complex conjugate of the (i,j)-element of B. Then

$$\mathbf{v} A \bar{\mathbf{v}}^T = (\mathbf{v}A)\bar{\mathbf{v}}^T = \lambda \mathbf{v} \bar{\mathbf{v}}^T. \tag{1}$$

Also, $$\mathbf{v} \bar{A}^T \bar{\mathbf{v}}^T = \mathbf{v}(\overline{\mathbf{v}A})^T = \mathbf{v}(\overline{\lambda \mathbf{v}})^T = \bar{\lambda} \mathbf{v} \bar{\mathbf{v}}^T \tag{2}$$

But A is real, so $\bar{A} = A$; also A is symmetric, so $A^T = A$. Thus $\bar{A}^T = A^T = A$, and so (1) and (2) yield $\lambda \mathbf{v} \bar{\mathbf{v}}^T = \bar{\lambda} \mathbf{v} \bar{\mathbf{v}}^T$. Rearranging this gives

$$0 = (\lambda - \bar{\lambda}) \mathbf{v} \bar{\mathbf{v}}^T$$
$$= (\lambda - \bar{\lambda})(v_1 \bar{v}_1 + \ldots + v_n \bar{v}_n) \quad \text{where } \bar{v}_i = \text{complex conjugate of } v_i$$
$$= (\lambda - \bar{\lambda})(|v_1|^2 + \ldots + |v_n|^2).$$

But $\mathbf{v} \neq \mathbf{0}$, and hence at least one of the v_is is non-zero. It follows that $\lambda - \bar{\lambda} = 0$, and hence that $\lambda = \bar{\lambda}$. Thus λ is real.

THEOREM 8.3.2   The eigenvectors corresponding to distinct eigenvalues of an $n \times n$ real symmetric matrix A are orthogonal.

Proof   Let $\lambda_1 \neq \lambda_2$ be two eigenvalues of A, and let $\mathbf{v}_1, \mathbf{v}_2$ be eigenvectors of A corresponding to λ_1, λ_2 respectively. Then

$$\mathbf{v}_1 A = \lambda_1 \mathbf{v}_1, \qquad \mathbf{v}_2 A = \lambda_2 \mathbf{v}_2.$$

Hence $(\mathbf{v}_1 A)\mathbf{v}_2^T = \lambda_1 \mathbf{v}_1 \mathbf{v}_2^T,$

and $\mathbf{v}_1(A\mathbf{v}_2^T) = \mathbf{v}_1(A^T \mathbf{v}_2^T) = \mathbf{v}_1(\mathbf{v}_2 A)^T = \lambda_2 \mathbf{v}_1 \mathbf{v}_2^T,$ since A is symmetric.

But $(\mathbf{v}_1 A)\mathbf{v}_2^T = \mathbf{v}_1(A\mathbf{v}_2^T),$ so $\lambda_1 \mathbf{v}_1 \mathbf{v}_2^T = \lambda_2 \mathbf{v}_1 \mathbf{v}_2^T.$

Rearranging this gives

$$0 = (\lambda_1 - \lambda_2)\mathbf{v}_1 \mathbf{v}_2^T,$$

which implies that $\mathbf{v}_1 \mathbf{v}_2^T = 0$ (since $\lambda_1 \neq \lambda_2$). Thus $\mathbf{v}_1 \cdot \mathbf{v}_2 = 0$ and $\mathbf{v}_1, \mathbf{v}_2$ are orthogonal.

THEOREM 8.3.3 Let $A = [a_{ij}]$ be a 2×2 real symmetric matrix. Then there is an orthogonal matrix P such that $P^T A P$ is a diagonal matrix.

Proof Let λ be an eigenvalue of A, and let $\mathbf{u} = (u_1, u_2)$ be a unit eigenvector of A corresponding to λ. (We can find such by taking any corresponding eigenvector $\mathbf{u}'$ and putting $\mathbf{u} = \mathbf{u}'/|\mathbf{u}'|$.) Choose $\mathbf{v} = (v_1, v_2)$ to be a unit vector such that $\mathbf{u} \cdot \mathbf{v} = 0$; we saw in section 1.3 how to do this.

Put
$$P = \begin{bmatrix} u_1 & v_1 \\ u_2 & v_2 \end{bmatrix}.$$

It is easy to check that P is orthogonal, and

$$P^T A P = \begin{bmatrix} u_1 & u_2 \\ v_1 & v_2 \end{bmatrix} \begin{bmatrix} a_{11} & a_{12} \\ a_{21} & a_{22} \end{bmatrix} \begin{bmatrix} u_1 & v_1 \\ u_2 & v_2 \end{bmatrix}$$

$$= \begin{bmatrix} u_1 a_{11} + u_2 a_{21} & u_1 a_{12} + u_2 a_{22} \\ v_1 a_{11} + v_2 a_{21} & v_1 a_{12} + v_2 a_{22} \end{bmatrix} \begin{bmatrix} u_1 & v_1 \\ u_2 & v_2 \end{bmatrix}$$

$$= \begin{bmatrix} \lambda u_1 & \lambda u_2 \\ w_1 & w_2 \end{bmatrix} \begin{bmatrix} u_1 & v_1 \\ u_2 & v_2 \end{bmatrix} \quad \text{where } (v_1 a_{11} + v_2 a_{21}, v_1 a_{12} + v_2 a_{22})$$
$$= (w_1, w_2) = \mathbf{w} \text{ (say)} = \mathbf{v} A;$$
$$\text{since } (u_1 a_{11} + u_2 a_{21}, u_1 a_{12} + u_2 a_{22})$$
$$= \mathbf{u} A = \lambda(u_1, u_2).$$

$$= \begin{bmatrix} \lambda(u_1^2 + u_2^2) & \lambda(u_1 v_1 + u_2 v_2) \\ w_1 u_1 + w_2 u_2 & w_1 v_1 + w_2 v_2 \end{bmatrix}$$

$$= \begin{bmatrix} \lambda |\mathbf{u}| & \lambda \mathbf{u} \cdot \mathbf{v} \\ \mathbf{w} \cdot \mathbf{u} & \mathbf{w} \cdot \mathbf{v} \end{bmatrix} = \begin{bmatrix} \lambda & 0 \\ 0 & \mu \end{bmatrix},$$

since $|\mathbf{u}| = 1$, $\mathbf{u} \cdot \mathbf{v} = 0$, $\mathbf{w} \cdot \mathbf{v} = \mu \in \mathbb{R}$, and

$$\mathbf{w} \cdot \mathbf{u} = (\mathbf{v} A) \cdot \mathbf{u} = \mathbf{v} A \mathbf{u}^T = \mathbf{v} A^T \mathbf{u}^T = \mathbf{v}(\mathbf{u} A)^T = \lambda \mathbf{v} \mathbf{u}^T = \lambda \mathbf{v} \cdot \mathbf{u} = 0.$$

THEOREM 8.3.4 Let $A = [a_{ij}]$ be a 3×3 real symmetric matrix. Then there is an orthogonal matrix P such that $P^T A P$ is a diagonal matrix.

Proof Let λ be an eigenvalue of A, and let $\mathbf{u} = (u_1, u_2, u_3)$ be a unit eigenvector of A corresponding to λ. Choose $\mathbf{v} = (v_1, v_2, v_3)$, $\mathbf{w} = (w_1, w_2, w_3)$ to be unit vectors such that $\mathbf{u} \cdot \mathbf{v} = \mathbf{u} \cdot \mathbf{w} = \mathbf{v} \cdot \mathbf{w} = 0$.

Put
$$P_1 = \begin{bmatrix} u_1 & v_1 & w_1 \\ u_2 & v_2 & w_2 \\ u_3 & v_3 & w_3 \end{bmatrix}.$$

Then P_1 is orthogonal and

$$P_1^T A P_1 = \begin{bmatrix} u_1 & u_2 & u_3 \\ v_1 & v_2 & v_3 \\ w_1 & w_2 & w_3 \end{bmatrix} \begin{bmatrix} a_{11} & a_{12} & a_{13} \\ a_{21} & a_{22} & a_{23} \\ a_{31} & a_{32} & a_{33} \end{bmatrix} \begin{bmatrix} u_1 & v_1 & w_1 \\ u_2 & v_2 & w_2 \\ u_3 & v_3 & w_3 \end{bmatrix}$$

$$= \begin{bmatrix} \lambda u_1 & \lambda u_2 & \lambda u_3 \\ x_1 & x_2 & x_3 \\ y_1 & y_2 & y_3 \end{bmatrix} \begin{bmatrix} u_1 & v_1 & w_1 \\ u_2 & v_2 & w_2 \\ u_3 & v_3 & w_3 \end{bmatrix} \quad \text{where } (x_1, x_2, x_3) = \mathbf{v}A = \mathbf{x} \text{ say,}$$
$$\text{and } (y_1, y_2, y_3) = \mathbf{w}A = \mathbf{y} \text{ say.}$$

$$= \begin{bmatrix} \lambda|\mathbf{u}| & \lambda \mathbf{u} \cdot \mathbf{v} & \lambda \mathbf{u} \cdot \mathbf{w} \\ \mathbf{x} \cdot \mathbf{u} & \mathbf{x} \cdot \mathbf{v} & \mathbf{x} \cdot \mathbf{w} \\ \mathbf{y} \cdot \mathbf{u} & \mathbf{y} \cdot \mathbf{v} & \mathbf{y} \cdot \mathbf{w} \end{bmatrix} = \begin{bmatrix} \lambda & \vdots & 0 & 0 \\ \cdots & \cdots & \cdots & \cdots \\ 0 & \vdots & & \\ 0 & \vdots & & B \end{bmatrix}$$

as in Theorem 8.3.3, where $B = \begin{bmatrix} \mathbf{x} \cdot \mathbf{v} & \mathbf{x} \cdot \mathbf{w} \\ \mathbf{y} \cdot \mathbf{v} & \mathbf{y} \cdot \mathbf{w} \end{bmatrix}$.

Now $\mathbf{x} \cdot \mathbf{w} = (\mathbf{v}A) \cdot \mathbf{w} = \mathbf{v}A\mathbf{w}^T = \mathbf{v}A^T\mathbf{w}^T = \mathbf{v}(\mathbf{w}A)^T = \mathbf{v} \cdot (\mathbf{w}A) = \mathbf{v} \cdot \mathbf{y} = \mathbf{y} \cdot \mathbf{v}$.

Thus B is a 2×2 real symmetric matrix. By Theorem 8.3.3, there is an orthogonal matrix P_2 such that $P_2^T B P_2$ is diagonal; put

$$P_2^T B P_2 = \begin{bmatrix} \mu & 0 \\ 0 & \gamma \end{bmatrix}, \qquad P_3 = \begin{bmatrix} 1 & \vdots & 0 & 0 \\ \cdots & \cdots & \cdots & \cdots \\ 0 & \vdots & & \\ 0 & \vdots & & P_2 \end{bmatrix}.$$

Then
$(P_1 P_3)^T A (P_1 P_3) = P_3^T (P_1^T A P_1) P_3$

$$= \begin{bmatrix} 1 & \vdots & 0 & 0 \\ \cdots & \cdots & \cdots & \cdots \\ 0 & \vdots & & \\ 0 & \vdots & & P_2^T \end{bmatrix} \begin{bmatrix} \lambda & \vdots & 0 & 0 \\ \cdots & \cdots & \cdots & \cdots \\ 0 & \vdots & & \\ 0 & \vdots & & B \end{bmatrix} \begin{bmatrix} 1 & \vdots & 0 & 0 \\ \cdots & \cdots & \cdots & \cdots \\ 0 & \vdots & & \\ 0 & \vdots & & P_2 \end{bmatrix}$$

$$= \begin{bmatrix} \lambda & \vdots & 0 & 0 \\ \cdots & \cdots & \cdots & \cdots \\ 0 & \vdots & & \\ 0 & \vdots & & P_2^T A P_2 \end{bmatrix} = \begin{bmatrix} \lambda & 0 & 0 \\ 0 & \mu & 0 \\ 0 & 0 & \gamma \end{bmatrix}.$$

Since $P_1 P_3$ is orthogonal, we may put $P = P_1 P_3$.

Theorems 8.3.3 and 8.3.4 are, of course, only existence results; we now have to consider how to find the matrix P in practice. The answer is given by the following result.

THEOREM 8.3.5 Let $\lambda_1, \lambda_2, \lambda_3$ be the eigenvalues of the 3×3 real symmetric matrix A with their proper multiplicities.

If $\quad \mathbf{u} = (u_1, u_2, u_3), \quad \mathbf{v} = (v_1, v_2, v_3), \quad \mathbf{w} = (w_1, w_2, w_3)$

are corresponding unit eigenvectors of A which are mutually orthogonal,

then $P = \begin{bmatrix} u_1 & v_1 & w_1 \\ u_2 & v_2 & w_2 \\ u_3 & v_3 & w_3 \end{bmatrix}$ is orthogonal, and $P^T A P = \text{diag}(\lambda_1, \lambda_2, \lambda_3)$.

Proof It is straightforward to check that P is orthogonal. Also,

$P^T A P = \begin{bmatrix} \lambda_1 \mathbf{u} \cdot \mathbf{u} & \lambda_2 \mathbf{u} \cdot \mathbf{v} & \lambda_3 \mathbf{u} \cdot \mathbf{w} \\ \lambda_1 \mathbf{v} \cdot \mathbf{u} & \lambda_2 \mathbf{v} \cdot \mathbf{v} & \lambda_3 \mathbf{v} \cdot \mathbf{w} \\ \lambda_1 \mathbf{w} \cdot \mathbf{u} & \lambda_2 \mathbf{w} \cdot \mathbf{v} & \lambda_3 \mathbf{w} \cdot \mathbf{w} \end{bmatrix}$ by a calculation as in the previous two results.

$= \begin{bmatrix} \lambda_1 & 0 & 0 \\ 0 & \lambda_2 & 0 \\ 0 & 0 & \lambda_3 \end{bmatrix}$ since $\mathbf{u}, \mathbf{v}, \mathbf{w}$ are mutually orthogonal unit vectors.

Thus, given a 3×3 real symmetric matrix A to reduce to diagonal form, we

(a) find the eigenvalues of A, then
(b) find mutually orthogonal unit eigenvectors of A corresponding to the eigenvalues.

Examples 8.3

There are three cases that can arise; we will illustrate each of them.

1. *Case 1: the eigenvalues $\lambda_1, \lambda_2, \lambda_3$ of A are all distinct.*

   Let $\mathbf{u}, \mathbf{v}, \mathbf{w}$ be *any* unit eigenvectors corresponding to $\lambda_1, \lambda_2, \lambda_3$. Then they are mutually orthogonal, by Theorem 8.3.2, so the required matrix is given by Theorem 8.3.5.

   Let $\quad A = \begin{bmatrix} 7 & -2 & -2 \\ -2 & 1 & 4 \\ -2 & 4 & 1 \end{bmatrix}$.

   Then $\quad \det(A - \lambda I_3) = \begin{vmatrix} 7-\lambda & -2 & -2 \\ -2 & 1-\lambda & 4 \\ -2 & 4 & 1-\lambda \end{vmatrix}$

$$= (7-\lambda)((1-\lambda)^2 - 16) + 2(2(\lambda-1)+8) - 2(-8 - 2(\lambda-1))$$
$$= (7-\lambda)(-3-\lambda)(5-\lambda) + 8(3+\lambda)$$
$$= (\lambda+3)(-35 + 12\lambda - \lambda^2 + 8)$$
$$= (\lambda+3)(\lambda-9)(3-\lambda).$$

Therefore the eigenvalues of A are 3, -3 and 9. Let $\mathbf{u}' = (u_1, u_2, u_3)$ be an eigenvector corresponding to the eigenvalue 9. Then $A\mathbf{u}' = 9\mathbf{u}'$, so

$$0 = (A - 9I_3)\mathbf{u}' = \begin{bmatrix} -2 & -2 & -2 \\ -2 & -8 & 4 \\ -2 & 4 & -8 \end{bmatrix} \begin{bmatrix} u_1 \\ u_2 \\ u_3 \end{bmatrix}.$$

Solving this system of linear equations gives

$$u_3 = \alpha \text{ (an arbitrary real number)}, \quad u_2 = \alpha, \quad u_1 = -2\alpha.$$

Hence $\mathbf{u}' = \alpha(-2, 1, 1)$, so a *particular* eigenvector corresponding to eigenvalue 9 is $(-2, 1, 1)$. Similarly we can find the eigenvectors corresponding to eigenvalues 3 and -3. Particular eigenvectors turn out to be $\mathbf{v}' = (1, 1, 1)$, $\mathbf{w}' = (0, -1, 1)$ corresponding to $\lambda = 3, -3$ respectively.

Now we want *unit* eigenvectors, so we choose

$$\mathbf{u} = \mathbf{u}'/|\mathbf{u}'| = (-2/\sqrt{6}, 1/\sqrt{6}, 1/\sqrt{6}),$$
$$\mathbf{v} = \mathbf{v}'/|\mathbf{v}'| = (1/\sqrt{3}, 1/\sqrt{3}, 1/\sqrt{3}),$$
$$\mathbf{w} = \mathbf{w}'/|\mathbf{w}'| = (0, -1/\sqrt{2}, 1/\sqrt{2}).$$

Hence a suitable orthogonal matrix is

$$P = \begin{bmatrix} -2/\sqrt{6} & 1/\sqrt{3} & 0 \\ 1/\sqrt{6} & 1/\sqrt{3} & -1/\sqrt{2} \\ 1/\sqrt{6} & 1/\sqrt{3} & 1/\sqrt{2} \end{bmatrix},$$

and it can be checked that

$$P^T A P = \begin{bmatrix} 9 & 0 & 0 \\ 0 & 3 & 0 \\ 0 & 0 & -3 \end{bmatrix}.$$

2. *Case 2: the eigenvalues of A are $\lambda_1, \lambda_1, \lambda_2$ (where $\lambda_1 \neq \lambda_2$).*

By Theorem 8.3.2 *any* eigenvector corresponding to λ_1 and *any* eigenvector corresponding to λ_2 are orthogonal. We thus have to find two orthogonal eigenvectors corresponding to λ_1. These certainly exist, though we shall relegate the proof of this non-trivial fact to a set of exercises.

We find two orthogonal eigenvectors $\mathbf{u}', \mathbf{v}'$ corresponding to λ_1 as follows:

(a) find two eigenvectors $\mathbf{u}_1, \mathbf{v}_1$ corresponding to λ_1 such that neither is a scalar multiple of the other;

(b) put $\mathbf{u}' = \mathbf{u}_1$, $\mathbf{v}' = \mathbf{v}_1 - \dfrac{\mathbf{u}_1 \cdot \mathbf{v}_1}{|\mathbf{u}_1|^2}\mathbf{u}_1$.

You should check that $\mathbf{u}'$ and $\mathbf{v}'$ are indeed orthogonal.

Let
$$A = \begin{bmatrix} 2 & 1 & 1 \\ 1 & 2 & 1 \\ 1 & 1 & 2 \end{bmatrix}.$$

Then the eigenvalues of A are $1, 4$. An eigenvector corresponding to the eigenvalue 4 is $\mathbf{w}' = (1, 1, 1)$. The eigenvectors corresponding to the eigenvalue 1 are precisely the vectors of the form $(\alpha, \beta, -\alpha - \beta)$ for any real numbers α, β (not both zero). So, in particular, we can put

$\alpha = 1, \beta = -1$ to give $\mathbf{u}_1 = (1, -1, 0)$

$\alpha = 1, \beta = 0$ to give $\mathbf{v}_1 = (1, 0, -1)$.

Clearly, $\mathbf{u}_1$ and $\mathbf{v}_1$ are not scalar multiples of each other. Now put

$\mathbf{u}' = \mathbf{u}_1 = (1, -1, 0),$

$\mathbf{v}' = \mathbf{v}_1 - \dfrac{\mathbf{u}_1 \cdot \mathbf{v}_1}{|\mathbf{u}_1|^2}\mathbf{u}_1 = (1, 0, -1) - \tfrac{1}{2}(1, -1, 0) = (\tfrac{1}{2}, \tfrac{1}{2}, -1).$

Then $\mathbf{u}', \mathbf{v}'$ are orthogonal eigenvectors corresponding to the eigenvalue 1. We can therefore take

$$\mathbf{u} = \mathbf{u}'/|\mathbf{u}'| = (1/\sqrt{2}, -1/\sqrt{2}, 0),$$
$$\mathbf{v} = \mathbf{v}'/|\mathbf{v}'| = (1/\sqrt{6}, 1/\sqrt{6}, -2/\sqrt{6}),$$
and $\quad \mathbf{w} = \mathbf{w}'/|\mathbf{w}'| = (1/\sqrt{3}, 1/\sqrt{3}, 1/\sqrt{3}).$

Hence, a suitable orthogonal matrix is
$$P = \begin{bmatrix} 1/\sqrt{2} & 1/\sqrt{6} & 1/\sqrt{3} \\ -1/\sqrt{2} & 1/\sqrt{6} & 1/\sqrt{3} \\ 0 & -2/\sqrt{6} & 1/\sqrt{3} \end{bmatrix}.$$

It can be checked that
$$P^T A P = \begin{bmatrix} 1 & 0 & 0 \\ 0 & 1 & 0 \\ 0 & 0 & 4 \end{bmatrix}.$$

3. *Case 3: the eigenvalues of A are all equal, to λ say.*

   In this case there is an orthogonal matrix P such that $P^T AP = \lambda I_3$ (by Theorem 8.3.5). Pre-multiplying both sides of this equation by P and post-multiplying both sides by P^T gives $A = \lambda PP^T = \lambda I_3$; that is, A is already diagonal. A suitable orthogonal matrix is, therefore, I_3.

   Note: The corresponding results hold for 2×2 real symmetric matrices. Why not try filling in the details?

EXERCISES 8.3

1. Recall that a square matrix A is called antisymmetric if $A^T = -A$. Deduce that the eigenvalues of a real antisymmetric matrix are all imaginary or zero. (*Hint:* Look at the proof of Theorem 8.3.1.)

2. For each of the following matrices A, find an orthogonal matrix P such that $P^T AP$ is in diagonal form.

   (a) $\begin{bmatrix} 2 & -1 \\ -1 & 2 \end{bmatrix}$, (b) $\begin{bmatrix} 3 & 1 \\ 1 & 3 \end{bmatrix}$, (c) $\begin{bmatrix} 0 & 1 & 1 \\ 1 & 0 & 1 \\ 1 & 1 & 0 \end{bmatrix}$,

   (d) $\begin{bmatrix} 2 & -3 & 3 \\ -3 & 10 & -9 \\ 3 & -9 & 10 \end{bmatrix}$, (e) $\begin{bmatrix} 2 & 1 & 1 \\ 1 & 2 & 1 \\ 1 & 1 & 2 \end{bmatrix}$, (f) $\begin{bmatrix} 0 & 1 & 0 \\ 1 & 0 & 0 \\ 0 & 0 & 2 \end{bmatrix}$,

   (g) $\begin{bmatrix} 1 & 0 & 0 \\ 0 & -2 & 4 \\ 0 & 4 & -2 \end{bmatrix}$.

3. Let P be an orthogonal matrix such that $P^T AP = \text{diag}(\lambda_1, \ldots, \lambda_n)$, where A is a real symmetric matrix. Prove that $\lambda_1, \ldots, \lambda_n$ are the eigenvalues of A with their proper multiplicities.

4. Let $P = \begin{bmatrix} u_1 & v_1 & w_1 \\ u_2 & v_2 & w_2 \\ u_3 & v_3 & w_3 \end{bmatrix}$ be an orthogonal matrix such that $P^T AP = \text{diag}(\lambda_1, \lambda_2, \lambda_3)$ where A is a 3×3 real symmetric matrix. By the previous exercise, $\lambda_1, \lambda_2, \lambda_3$ are the eigenvalues of A with their proper multiplicities. Prove that $\mathbf{u} = (u_1, u_2, u_3)$, $\mathbf{v} = (v_1, v_2, v_3)$, $\mathbf{w} = (w_1, w_2, w_3)$ are corresponding unit eigenvectors.

8.4 CLASSIFICATION OF CONICS

Throughout we shall assume that we have a fixed right-handed rectangular coordinate system to which our equations refer. A *conic* is a set of points (x, y) in the plane satisfying an equation of the form

$$ax^2 + 2hxy + by^2 + 2gx + 2fy + c = 0 \qquad (1)$$

where a, b, h, g, f, c are real numbers. The equation (1) is called the *equation of the conic*. If we put $A = \begin{bmatrix} a & h \\ h & b \end{bmatrix}$, $\mathbf{d} = (g, f)$, $\mathbf{x} = (x, y)$, then (1) can be rewritten as

$$\mathbf{x} A \mathbf{x}^T + 2\mathbf{d}\mathbf{x}^T + c = 0 \qquad (2)$$

We shall investigate the different types of conic that can arise corresponding to different values of the coefficients in (1). Let us consider first the effect of rotating the axes. Consider the Euclidean transformation R_θ described in section 8.2. Then $R_\theta(\mathbf{x}) = \mathbf{x}P = \mathbf{y}$ (say), where P is orthogonal and $\det P = 1$, by Corollary 8.2.3. Since $P^{-1} = P^T$, $\mathbf{y}P^T = (\mathbf{x}P)P^{-1} = \mathbf{x}$. Thus,

$\mathbf{x}$ lies on the conic (2) $\Leftrightarrow \mathbf{x}A\mathbf{x}^T + 2\mathbf{d}\mathbf{x}^T + c = 0$

$\Leftrightarrow \mathbf{y}P^T A(\mathbf{y}P^T)^T + 2\mathbf{d}(\mathbf{y}P^T)^T + c = 0$

$\Leftrightarrow \mathbf{y}$ lies on the conic
$\mathbf{y}(P^T A P)\mathbf{y}^T + 2\mathbf{d}P\mathbf{y}^T + c = 0$.

Now the 2×2 analogue of Theorem 8.3.5 tells us that we can find an orthogonal matrix P such that $P^T AP = \text{diag}(\lambda_1, \lambda_2)$, where λ_1, λ_2 are the eigenvalues of A. Furthermore, we can find such a P with $\det P = 1$. For, $\det P = \pm 1$ by Lemma 8.1.1(a), and if $\det P = -1$ then simply change the sign of every element in the first column of P. (Note that the resulting matrix will still be orthogonal and will have determinant 1.) As any orthogonal matrix P with $\det P = 1$ is associated with a rotation, we have proved the following result.

THEOREM 8.4.1 There is a rotation R_θ of the plane which transforms the conic (1) into the conic

$$\lambda_1 x^2 + \lambda_2 y^2 + 2\phi x + 2\psi y + c = 0. \qquad (3)$$

Let us analyse the different cases which can arise.

Case 1 $\lambda_1, \lambda_2 \neq 0$

Rearrange (3) to give

$$\lambda_1 \left(x + \frac{\phi}{\lambda_1} \right)^2 + \lambda_2 \left(y + \frac{\psi}{\lambda_2} \right)^2 = \frac{\phi^2}{\lambda_1} + \frac{\psi^2}{\lambda_2} - c.$$

Making the translation

$$x_1 = x + \frac{\phi}{\lambda_1}, \qquad y_1 = y + \frac{\psi}{\lambda_2}$$

transforms the equation of the conic into

$$\lambda_1 x_1^2 + \lambda_2 y_1^2 = p, \qquad \text{where } p = \frac{\phi^2}{\lambda_1} + \frac{\psi^2}{\lambda_2} - c.$$

If (a) $\lambda_1 > 0, \lambda_2 > 0, p > 0$ this represents an *ellipse*;
(b) $\lambda_1 > 0, \lambda_2 < 0, p > 0$ this represents a *hyperbola*;
(c) $\lambda_1 < 0, \lambda_2 < 0, p > 0$ this represents the *empty set*;
(d) $\lambda_1 > 0, \lambda_2 > 0, p = 0$ this represents the *origin*;
(e) $\lambda_1 > 0, \lambda_2 < 0, p = 0$ this represents a *pair of straight lines*
$$\sqrt{\lambda_1} x_1 \pm \sqrt{-\lambda_2} y_1 = 0.$$

Case 2   One of λ_1, λ_2 is zero; say $\lambda_2 = 0$.

Rearrange (3) to give

$$\lambda_1 \left(x + \frac{\phi}{\lambda_1} \right)^2 + 2\psi y = \frac{\phi^2}{\lambda_1} - c.$$

Making the translation

$$x_1 = x + \frac{\phi}{\lambda_1}$$

transforms the equation of the conic into

$$\lambda_1 x_1^2 + 2\psi y = p.$$

If $\psi \neq 0$, putting $y_1 = y - \frac{p}{2\psi}$, $\mu = 2\psi$ gives $\lambda_1 x_1^2 + \mu y_1 = 0$.

If (a) $\psi \neq 0, \lambda_1 \neq 0$ this represents a *parabola*;
(b) $\psi \neq 0, \lambda_1 = 0$ this represents the *x-axis*;
(c) $\psi = 0; \lambda_1, p$ same sign this represents the *parallel lines*
$$y_1 = \pm\sqrt{(p/\lambda_1)}$$
(d) $\psi = 0; \lambda_1, p$ opposite sign this represents the *empty set*.

All possibilities can be reduced to one of the above cases by multiplying both sides by -1, or by interchanging x_1 and y_1.

When the linear terms can be eliminated (that is, in every case apart from case 2(a), the parabola) the conic is called *central*, and the form $\lambda_1 x^2 + \lambda_2 y^2 = p$ is called the *normal form*. The *centre* is defined to be the point which moves to the origin under the Euclidean transformations which reduce the conic to normal form. In fact, only the translation is involved, as the rotation fixes the origin. Hence the centre is the point

$$(-\phi/\lambda_1, -\psi/\lambda_2), \quad \text{where } (\phi, \psi) = \mathbf{d}P.$$

The conics described in case 1(c), (d), (e) and case 2(b), (c), (d) are called *degenerate*. The only non-degenerate cases are, therefore, the ellipse, the hyperbola, and the parabola.

Example 8.4

Consider the conic

$$3x^2 + 3y^2 + 2xy + 6\sqrt{2}x + 6\sqrt{2}y + 3 = 0.$$

Then $\quad A = \begin{bmatrix} 3 & 1 \\ 1 & 3 \end{bmatrix}, \mathbf{d} = (3\sqrt{2}, 3\sqrt{2}), c = 3.$

The eigenvalues of A are 2, 4, and corresponding unit eigenvectors are $(1/\sqrt{2}, -1/\sqrt{2})$ and $(1/\sqrt{2}, 1/\sqrt{2})$ respectively.

Hence $\quad P = \begin{bmatrix} 1/\sqrt{2} & 1/\sqrt{2} \\ -1/\sqrt{2} & 1/\sqrt{2} \end{bmatrix} \quad$ and $\quad (\phi, \psi) = \mathbf{d}P = (0, 6).$

Thus, $\quad p = 0 + (36/4) - 3 = 6.$

The conic is, therefore, an ellipse with centre $(0, -3/2)$ and normal form $2x^2 + 4y^2 = 6$.

EXERCISES 8.4

1 Show that the matrix $A = \begin{bmatrix} 1/\sqrt{2} & -1/\sqrt{2} \\ 1/\sqrt{2} & 1/\sqrt{2} \end{bmatrix}$ is orthogonal. A conic C has equation $x^2 - 2xy + y^2 + \sqrt{2}x - \sqrt{2}y - 4 = 0$. Find the equation of the conic into which C is transformed under the linear transformation with matrix A (with respect to standard bases). Hence, or otherwise, show that the conic C is a line-pair.

2 Classify the following conics, and, where possible, find their centres.
   (a) $-6x^2 + 24xy + y^2 + 36x - 2y - 2 = 0$;
   (b) $x^2 + y^2 + 4x - 6y + 14 = 0$;
   (c) $3\sqrt{2}x^2 - 2\sqrt{2}xy + 3\sqrt{2}y^2 - 20x + 12y + 16\sqrt{2} = 0$;
   (d) $4\sqrt{5}x^2 - 4\sqrt{5}xy + \sqrt{5}y^2 + 80x - 40y + 75\sqrt{5} = 0$;
   (e) $9x^2 - 24xy + 16y^2 + 40x + 30y - 100 = 0$.

8.5 CLASSIFICATION OF QUADRICS

As in section 8.4 we assume that we have a fixed right-handed rectangular coordinate system to which our equations refer. A *quadric* is a set of points (x, y, z) in three-dimensional space satisfying an equation of the form

$$ax^2 + by^2 + cz^2 + 2fyz + 2gzx + 2hxy + 2px + 2qy + 2rz + k = 0 \qquad (1)$$

where $a, b, c, f, g, h, p, q, r, k$ are real numbers. If we put

$$A = \begin{bmatrix} a & h & g \\ h & b & f \\ g & f & c \end{bmatrix}, \qquad \mathbf{d} = (p, q, r), \qquad \mathbf{x} = (x, y, z),$$

then (1) can be rewritten as

$$\mathbf{x}A\mathbf{x}^T + 2\mathbf{d}\mathbf{x}^T + k = 0 \qquad (2)$$

As in section 8.4 there is a rotation R_θ given by $R_\theta(\mathbf{x}) = \mathbf{x}P = \mathbf{y}$ (say), where P is orthogonal and $\det P = 1$, which transforms the quadric into

$$\mathbf{y}(P^T A P)\mathbf{y} + 2\mathbf{d}P\mathbf{y}^T + k = 0;$$

that is, by choosing P such that $P^T A P = \operatorname{diag}(\lambda_1, \lambda_2, \lambda_3)$, where $\lambda_1, \lambda_2, \lambda_3$ are the eigenvalues of A, we obtain

THEOREM 8.6.1 There is a rotation R_θ of $\mathbb{R}^3$ which transforms the quadric (1) into the quadric

$$\lambda_1 x^2 + \lambda_2 y^2 + \lambda_3 z^2 + 2\phi x + 2\psi y + 2\theta z + k = 0 \qquad (3)$$

Again we analyse the different cases which can arise.

Case 1 $\lambda_1, \lambda_2, \lambda_3 \neq 0$.

Rearrange (3) to give

$$\lambda_1\left(x + \frac{\phi}{\lambda_1}\right)^2 + \lambda_2\left(y + \frac{\psi}{\lambda_2}\right)^2 + \lambda_3\left(z + \frac{\theta}{\lambda_3}\right)^2 = \frac{\phi^2}{\lambda_1} + \frac{\psi^2}{\lambda_2} + \frac{\theta^2}{\lambda_3} - k.$$

Making the translation

$$x_1 = x + \frac{\phi}{\lambda_1}, \qquad y_1 = y + \frac{\psi}{\lambda_2}, \qquad z_1 = z + \frac{\theta}{\lambda_3}$$

transforms the equation of the quadric into

$$\lambda_1 x_1^2 + \lambda_2 y_1^2 + \lambda_3 z_1^2 = s, \qquad \text{where} \qquad s = \frac{\phi^2}{\lambda_1} + \frac{\psi^2}{\lambda_2} + \frac{\theta^2}{\lambda_3} - k.$$

(a) If $\lambda_1 > 0, \lambda_2 > 0, \lambda_3 > 0, s > 0$ this represents an *ellipsoid* (Fig. 8.3).

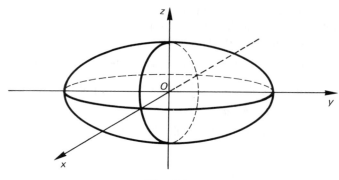

Fig. 8.3

(b) If $\lambda_1 > 0, \lambda_2 > 0, \lambda_3 < 0, s > 0$ this represents a *hyperboloid of one sheet* (Fig. 8.4).

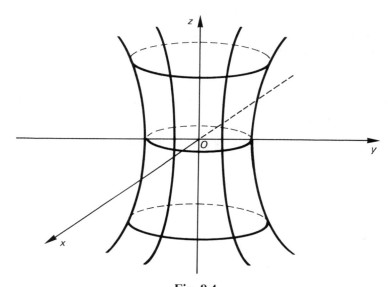

Fig. 8.4

(c) If $\lambda_1 > 0, \lambda_2 < 0, \lambda_3 < 0, s > 0$ this represents a *hyperboloid of two sheets* (Fig. 8.5).
(d) If $\lambda_1 < 0, \lambda_2 < 0, \lambda_3 < 0, s > 0$ this represents the *empty set*.
(e) If $\lambda_1 > 0, \lambda_2 > 0, \lambda_3 > 0, s = 0$ this represents the *origin*.
(f) If $\lambda_1 > 0, \lambda_2 > 0, \lambda_3 < 0, s = 0$ this represents a *cone* (Fig. 8.6).

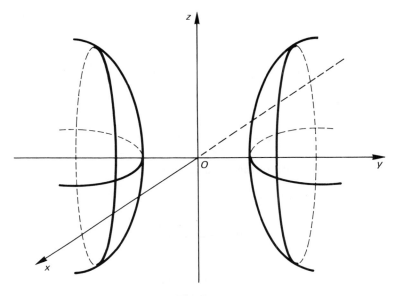

Fig. 8.5

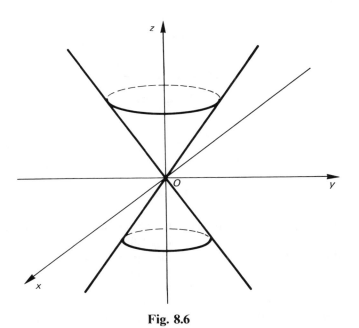

Fig. 8.6

Case 2 $\lambda_1, \lambda_2 \neq 0, \lambda_3 = 0$.

Rearrange (3) to give

$$\lambda_1\left(x+\frac{\phi}{\lambda_1}\right)^2 + \lambda_2\left(y+\frac{\psi}{\lambda_2}\right)^2 + 2\theta z = \frac{\phi^2}{\lambda_1} + \frac{\psi^2}{\lambda_2} - k.$$

Making the translation

$$x_1 = x + \frac{\phi}{\lambda_1}, \qquad y_1 = y + \frac{\psi}{\lambda_2}$$

transforms the equation of the quadric into

$$\lambda_1 x_1^2 + \lambda_2 y_1^2 + 2\theta z = s, \qquad \text{where} \qquad s = \frac{\phi^2}{\lambda_1} + \frac{\psi^2}{\lambda_2} - k.$$

If $\theta \neq 0$, putting $z_1 = z - (s/2\theta)$, $\mu = 2\theta$ gives

$$\lambda_1 x_1^2 + \lambda_2 y_1^2 + \mu z_1 = 0.$$

(a) If $\theta \neq 0$; λ_1, λ_2 same sign, this represents an *elliptic paraboloid* (Fig. 8.7).

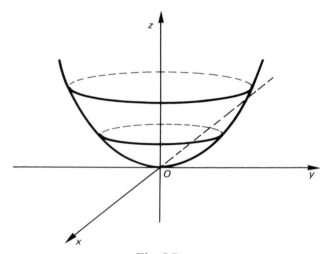

Fig. 8.7

(b) If $\theta \neq 0$; λ_1, λ_2 different sign, this represents a *hyperbolic paraboloid* (Fig. 8.8).
(c) If $\theta = 0$; $\lambda_1 > 0, \lambda_2 > 0, s > 0$, this represents an *elliptic cylinder* (Fig. 8.9).
(d) If $\theta = 0$; $\lambda_1 > 0, \lambda_2 < 0, s > 0$, this represents a *hyperbolic cylinder* (Fig. 8.10).
(e) If $\theta = 0$; $\lambda_1 > 0, \lambda_2 > 0, s = 0$, this represents the *z*-axis.

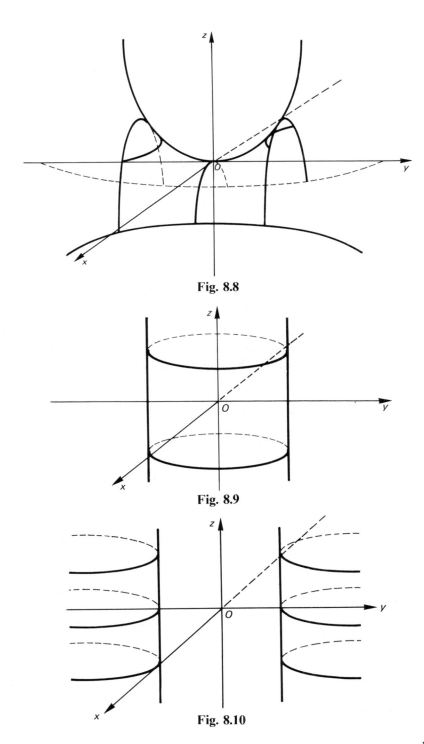

Fig. 8.8

Fig. 8.9

Fig. 8.10

(f) If $\theta = 0$; $\lambda_1 > 0, \lambda_2 < 0, s = 0$, this represents the *pair of planes*
$$\sqrt{\lambda_1} x_1 \pm \sqrt{-\lambda_2} y_1 = 0.$$
(g) If $\theta = 0$; $\lambda_1 > 0, \lambda_2 > 0, s < 0$, this represents the *empty set*.

Case 3 $\lambda_1 \neq 0, \lambda_2, \lambda_3 = 0$.

Rearrange (3) to give

$$\lambda_1 \left(x + \frac{\phi}{\lambda_1} \right)^2 + 2\psi y + 2\theta z + k = 0.$$

Making the translation $\qquad x_1 = x + \dfrac{\phi}{\lambda_1}$

transforms the equation of the quadric into

$$\lambda_1 x_1^2 + 2\psi y + 2\theta z + k - \frac{\phi^2}{\lambda_1} = 0.$$

If ψ or θ is non-zero, put $ty_1 = 2\psi y + 2\theta z + k - \dfrac{\phi^2}{\lambda_1}$. Notice that y_1 is in the (y, z)-plane, so by choosing t suitably, and z_1 to complete a right-handed rectangular coordinate system with x_1, y_1, the result can be obtained by means of a Euclidean transformation. The equation becomes

$$\lambda_1 x_1^2 + ty_1 = 0.$$

If $\psi = \theta = 0$, the equation becomes

$$\lambda_1 x_1^2 = s, \qquad \text{where } s = k - \frac{\phi^2}{\lambda_1}.$$

(a) If ψ is non-zero, this represents a *parabolic cylinder* (Fig. 8.11).
(b) If $\psi = \theta = 0, \lambda_1 > 0, s > 0$, this represents the *pair of parallel planes* $x_1 = \pm \sqrt{(s/\lambda_1)}$.
(c) If $\psi = \theta = 0, \lambda_1 > 0, s = 0$, this represents the (y, z)-plane.
(d) If $\psi = \theta = 0, \lambda_1 > 0, s < 0$, this represents the *empty set*.

All possibilities can be reduced to one of the above cases by multiplying both sides by -1 or by interchanging axes.

The quadrics considered in case 1 are called the *central* quadrics. As with the central conics the *centre* can be defined, and, analogously, is the point $(-\phi/\lambda_1, -\psi/\lambda_2, -\theta/\lambda_3)$, where $(\phi, \psi, \theta) = \mathbf{d}P$.

The quadrics described in case 1 (d), (e), (f), case 2 (c), (d), (e), (f), (g), and case 3 (a), (b), (c), (d) are called *degenerate*. The only non-degenerate quadrics are, therefore, the ellipsoid, the hyperboloid of one sheet, the hyperboloid of two sheets, the elliptic paraboloid, and the hyperbolic paraboloid.

ORTHOGONAL REDUCTION OF SYMMETRIC MATRICES

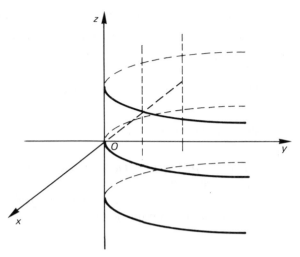

Fig. 8.11

EXERCISES 8.5

1. Let $P = \begin{bmatrix} 1/\sqrt{2} & -1/\sqrt{2} & 0 \\ 1/2 & 1/2 & 1/\sqrt{2} \\ 1/2 & 1/2 & -1/\sqrt{2} \end{bmatrix}$. Find the equation of the quadric Q_2 into which the quadric Q_1 with equation

$$2x^2 + 2y^2 + 3z^2 + 2xy + 2yz + 2zx = 1$$

is transformed by the linear transformation with matrix P (with respect to standard bases). Hence, or otherwise, identify the quadric Q_1.

2. Classify the following quadrics:
   (a) $2x^2 + 2y^2 + 2z^2 + 2xy + 2xz + 2yz = 1$;
   (b) $2x^2 + z^2 + 2xy + 6xz + 8yz = 0$.

SOLUTIONS AND HINTS FOR EXERCISES

Exercises 8.1

1. $\cos\theta = \dfrac{\mathbf{v}_1 \cdot \mathbf{v}_2}{|\mathbf{v}_1||\mathbf{v}_2|} = \dfrac{3}{\sqrt{30}\sqrt{6}} = 1/\sqrt{20}$.

Let $\mathbf{v} = (a, b, c, d)$. Then we want $\mathbf{v} \cdot \mathbf{v}_1 = \mathbf{v} \cdot \mathbf{v}_2 = \mathbf{v} \cdot \mathbf{v}_3 = 0$; that is,

$$a + 2b + 3c + 4d = 0,\ -b - c + 2d = 0,\ d = 0.$$

Thus, $a = b = -c$, and $\mathbf{v} = (c, c, -c, 0)$ where c is any non-zero real number.

2 Let $a\mathbf{v}_1 + b\mathbf{v}_2 + c\mathbf{v}_3 + d\mathbf{v}_4 = \mathbf{0}$. Taking scalar products of both sides with $\mathbf{v}_1$ gives $0 = \mathbf{0} \cdot \mathbf{v}_1 = a\mathbf{v}_1 \cdot \mathbf{v}_1 + b\mathbf{v}_2 \cdot \mathbf{v}_1 + c\mathbf{v}_3 \cdot \mathbf{v}_1 + d\mathbf{v}_4 \cdot \mathbf{v}_1 = a$. Similarly, taking the scalar product of each side with $\mathbf{v}_2, \mathbf{v}_3, \mathbf{v}_4$ in turn gives $b = c = d = 0$.

3 Let $A^T A = I$, $B^T B = I$. Then $(AB^T)^T A B^T = BA^T A B^T = BIB^T = BB^T = I$, by Corollary 8.1.2.

4 This is immediate from Theorem 8.1.4.

5 (a) $A = I^T A I$;
   (b) $A = P^T BP \Rightarrow PAP^T = PP^T BPP^T = IBI = B \Rightarrow B = (P^T)^T A P^T$ (and P^T is orthogonal by Corollary 8.1.2);
   (c) $A = P^T BP$, $B = Q^T CQ \Rightarrow A = P^T Q^T CQP = (QP)^T C(QP)$ (and QP is orthogonal by example 8.1.3).

Exercises 8.2

1 Let $\mathbf{v} = (x_1, y_1)$, $\mathbf{w} = (x_2, y_2)$. Then $\mathbf{v} - \mathbf{w} = (x_1 - x_2, y_1 - y_2)$,
$T(\mathbf{v}) - T(\mathbf{w}) = (x_1 - x_2, y_2 - y_1)$,
and $|\mathbf{v} - \mathbf{w}| = \sqrt{(x_1 - x_2)^2 + (y_1 - y_2)^2} = |T(\mathbf{v}) - T(\mathbf{w})|$.

2 Here $T(\mathbf{v}) - T(\mathbf{w})$
$= ((x_1 - x_2)\cos\theta - (y_1 - y_2)\sin\theta, (x_1 - x_2)\sin\theta + (y_1 - y_2)\cos\theta)$, so
$|T(\mathbf{v}) - T(\mathbf{w})| = \sqrt{(x_1 - x_2)^2 + (y_1 - y_2)^2}$ (using $\cos^2\theta + \sin^2\theta = 1$).

3 The required matrix is $A = \begin{bmatrix} \cos\theta & -\sin\theta & 0 \\ \sin\theta & \cos\theta & 0 \\ 0 & 0 & 1 \end{bmatrix}$, and $A^T A = I_3$.

4 Let $T: \mathbb{R}^n \to \mathbb{R}^n$ be a linear transformation, and let $\mathbf{e}_1, \ldots, \mathbf{e}_n$ be an orthonormal basis for $\mathbb{R}^n$. Suppose that T is Euclidean. Then, as in the proof of Theorem 8.1.4 ((b) implies (c)), T preserves scalar products. Hence

$$T(\mathbf{e}_i) \cdot T(\mathbf{e}_i) = \mathbf{e}_i \cdot \mathbf{e}_i = 1, \quad T(\mathbf{e}_i) \cdot T(\mathbf{e}_j) = \mathbf{e}_i \cdot \mathbf{e}_j = 0 \quad \text{if } i \neq j.$$

The fact that $T(\mathbf{e}_1), \ldots, T(\mathbf{e}_n)$ forms a basis for $\mathbb{R}^n$ follows from Theorem 8.2.1, Lemma 8.1.1, Theorem 6.3.1(d), and Theorem 6.2.2.

Now suppose that $T(\mathbf{e}_1), \ldots, T(\mathbf{e}_n)$ is an orthonormal basis for $\mathbb{R}^n$, and let $\mathbf{v}, \mathbf{w} \in \mathbb{R}^n$.

Put
$$\mathbf{v} = \sum_{i=1}^{n} \lambda_i \mathbf{e}_i, \quad \mathbf{w} = \sum_{i=1}^{n} \mu_i \mathbf{e}_i.$$

Then
$$T(\mathbf{v}) - T(\mathbf{w}) = \sum_{i=1}^{n} (\lambda_i - \mu_i) T(\mathbf{e}_i),$$

and so $|T(\mathbf{v}) - T(\mathbf{w})| = \sum_{i=1}^{n} \sum_{j=1}^{n} (\lambda_i - \mu_i)(\lambda_j - \mu_j) T(\mathbf{e}_i) \cdot T(\mathbf{e}_j)$

$$= \sum_{i=1}^{n} (\lambda_i - \mu_i)^2 = \sum_{i=1}^{n} \sum_{j=1}^{n} (\lambda_i - \mu_i)(\lambda_j - \mu_j) \mathbf{e}_i \cdot \mathbf{e}_j$$
$$= |\mathbf{v} - \mathbf{w}|.$$

5 Let $T: \mathbb{R}^n \to \mathbb{R}^n$ be a linear transformation, let $\mathbf{e}_1, \ldots, \mathbf{e}_n$ be an orthonormal basis for $\mathbb{R}^n$, and let A be the matrix associated with T with respect to this basis.

If $\mathbf{v} = \sum_{i=1}^{n} \lambda_i \mathbf{e}_i$, $\mathbf{w} = \sum_{i=1}^{n} \mu_i \mathbf{e}_i$, put $\boldsymbol{\lambda} = (\lambda_1, \ldots, \lambda_n)$, $\boldsymbol{\mu} = (\mu_1, \ldots, \mu_n)$,

so that $T(\mathbf{v}) = \boldsymbol{\lambda} A^T$, $T(\mathbf{w}) = \boldsymbol{\mu} A^T$. Then

T is Euclidean $\Leftrightarrow |\mathbf{v} - \mathbf{w}| = |T(\mathbf{v}) - T(\mathbf{w})|$ for all $\mathbf{v}, \mathbf{w} \in \mathbb{R}^n$

$\Leftrightarrow |\mathbf{v} - \mathbf{w}| = |(\boldsymbol{\lambda} - \boldsymbol{\mu}) A^T|$ for all $\mathbf{v}, \mathbf{w} \in \mathbb{R}^n$

$\Leftrightarrow |\mathbf{x}| = |\mathbf{x} A^T|$ for all $\mathbf{x} \in \mathbb{R}^n$

$\Leftrightarrow A^T$ is orthogonal (by Theorem 8.1.4)

$\Leftrightarrow A$ is orthogonal (by Corollary 8.1.3).

Exercises 8.3

1 Following the proof of Theorem 8.3.1 we find that $\bar{\lambda} = -\lambda$, from which the result follows easily.

2 (a) $\begin{bmatrix} 1/\sqrt{2} & 1/\sqrt{2} \\ -1/\sqrt{2} & 1/\sqrt{2} \end{bmatrix}$; (b) $\begin{bmatrix} 1/\sqrt{2} & 1/\sqrt{2} \\ 1/\sqrt{2} & -1/\sqrt{2} \end{bmatrix}$;

(c) $\begin{bmatrix} 1/\sqrt{2} & 1/\sqrt{6} & 1/\sqrt{3} \\ -1/\sqrt{2} & 1/\sqrt{6} & 1/\sqrt{3} \\ 0 & -2/\sqrt{6} & 1/\sqrt{3} \end{bmatrix}$; (d) $\begin{bmatrix} -3/\sqrt{10} & 3/\sqrt{190} & 1/\sqrt{19} \\ 0 & 3/\sqrt{190} & -3/\sqrt{19} \\ 1/\sqrt{10} & 3/\sqrt{190} & 6/\sqrt{19} \end{bmatrix}$;

(e) $\begin{bmatrix} 0 & -2/\sqrt{6} & 1/\sqrt{3} \\ 1/\sqrt{2} & 1/\sqrt{6} & 1/\sqrt{3} \\ -1/\sqrt{2} & 1/\sqrt{6} & 1/\sqrt{3} \end{bmatrix}$; (f) $\begin{bmatrix} 1/\sqrt{2} & 1/\sqrt{2} & 0 \\ 1/\sqrt{2} & -1/\sqrt{2} & 0 \\ 0 & 0 & 1 \end{bmatrix}$;

(g) $\begin{bmatrix} 1 & 0 & 0 \\ 0 & 1/\sqrt{2} & 1/\sqrt{2} \\ 0 & 1/\sqrt{2} & -1/\sqrt{2} \end{bmatrix}$.

3 Recall that $\lambda_1, \ldots, \lambda_n$ are all real. Suppose that μ is any real number. Then $P^T(A - \mu I)P = P^T AP - \mu P^T P = P^T AP - \mu I$. Taking determinants of both sides we have $\det(P^T(A - \mu I)P) = \det(P^T AP - \mu I)$; that is, $\det(P^T P) \det(A - \mu I) = \det(P^T AP - \mu I)$. But $\det(P^T P) = \det(I) = 1$, so

$$\det(A - \mu I) = \det(P^T A P - \mu I) = (\lambda_1 - \mu)\ldots(\lambda_n - \mu). \tag{1}$$

Hence, if λ is an indeterminate, then $\det(A - \lambda I)$ and $(\lambda_1 - \lambda)\ldots(\lambda_n - \lambda)$ are both polynomials of degree n which, by (1), take the same value for all real numbers $\mu = \lambda$. Hence, $\det(A - \lambda I) = (\lambda_1 - \lambda)\ldots(\lambda_n - \lambda)$, and the result follows.

4 The elements on the main diagonal of $P^T P$ are $\mathbf{u}\cdot\mathbf{u}, \mathbf{v}\cdot\mathbf{v}, \mathbf{w}\cdot\mathbf{w}$; since $P^T P = I$, then $\mathbf{u}, \mathbf{v}, \mathbf{w}$ must be unit vectors. Furthermore, if $P^T A P = \mathrm{diag}(\lambda_1, \lambda_2, \lambda_3)$ then $AP = P\,\mathrm{diag}(\lambda_1, \lambda_2, \lambda_3)$, and so $(A\mathbf{u}^T, A\mathbf{v}^T, A\mathbf{w}^T) = (\lambda_1 \mathbf{u}^T, \lambda_2 \mathbf{v}^T, \lambda_3 \mathbf{w}^T)$. It follows that $\mathbf{u}, \mathbf{v}, \mathbf{w}$ are eigenvectors for A.

Exercises 8.4

1 It is easy to check that $A^T A = I_2$. The linear transformation with matrix A is

$$T: \mathbb{R}^2 \to \mathbb{R}^2 \quad \text{where} \quad T((x,y)) = (x,y)A^T,$$

so $\quad (x', y')A = (x,y) = (x'/\sqrt{2} + y'/\sqrt{2}, -x'/\sqrt{2} + y'/\sqrt{2}).$

Hence the conic C_2 has equation

$$0 = (x'/\sqrt{2} + y'/\sqrt{2})^2 - 2(x'/\sqrt{2} + y'/\sqrt{2})(-x'/\sqrt{2} + y'/\sqrt{2})$$
$$+ (-x'/\sqrt{2} + y'/\sqrt{2})^2 + \sqrt{2}(x'/\sqrt{2} + y'/\sqrt{2})$$
$$- \sqrt{2}(-x'/\sqrt{2} + y'/\sqrt{2}) - 4$$
$$= 2x'^2 - 2x' - 4 = 2(x' - 2)(x' + 1),$$

which represents the line-pair $x' = -1$, $x' = 2$. Since A is orthogonal, T is euclidean, and so C represents a line-pair.

2 (a) $A = \begin{bmatrix} -6 & 12 \\ 12 & 1 \end{bmatrix}$, $\mathbf{d} = (18, -1)$, $c = -2$. The eigenvalues of A are 10, -15, and corresponding unit eigenvectors are $(3/5, 4/5)$, $(-4/5, 3/5)$ respectively.

Hence $P = \begin{bmatrix} 3/5 & -4/5 \\ 4/5 & 3/5 \end{bmatrix}$, $(\phi, \psi) = \mathbf{d}P = (10, -25)$.

Thus, $\lambda_1 > 0$, $\lambda_2 < 0$, $p < 0$; this is case 1(b), and the conic is a hyperbola with centre $(-1, -5/3)$.

(b) $A = \begin{bmatrix} 1 & 0 \\ 0 & 1 \end{bmatrix}$, $\mathbf{d} = (2, -3)$, $c = 14$. The eigenvalues of A are 1, 1, and corresponding unit eigenvectors are $(1, 0), (0, 1)$.

Hence $P = \begin{bmatrix} 1 & 0 \\ 0 & 1 \end{bmatrix}$, $(\phi, \psi) = (2, -3), p = -1$.

Thus, $\lambda_1 > 0$, $\lambda_2 > 0$, $p < 0$; this is case 1(c), and the conic is the empty set.

(c) $A = \begin{bmatrix} 3\sqrt{2} & -\sqrt{2} \\ -\sqrt{2} & 3\sqrt{2} \end{bmatrix}$, $\mathbf{d} = (-10, 6)$, $c = 16\sqrt{2}$. The eigenvalues of A are $2\sqrt{2}, 4\sqrt{2}$ and corresponding unit eigenvectors are $(1/\sqrt{2}, 1/\sqrt{2})$, $(1/\sqrt{2}, -1/\sqrt{2})$.

Hence $P = \begin{bmatrix} 1/\sqrt{2} & 1/\sqrt{2} \\ 1/\sqrt{2} & -1/\sqrt{2} \end{bmatrix}$,

$(\phi, \psi) = \mathbf{d}P = (-2\sqrt{2}, -8\sqrt{2}), p = 2\sqrt{2}$.

Thus, $\lambda_1 > 0$, $\lambda_2 > 0$, $p > 0$; this is case 1(a), and the conic is an ellipse with centre $(1, 2)$.

(d) $A = \begin{bmatrix} 4\sqrt{5} & -2\sqrt{5} \\ -2\sqrt{5} & \sqrt{5} \end{bmatrix}$, $\mathbf{d} = (40, -20)$, $c = 75\sqrt{5}$. The eigenvalues of A are $5\sqrt{5}, 0$ and corresponding unit eigenvectors are $(-2/\sqrt{5}, 1/\sqrt{5})$, $(1/\sqrt{5}, 2/\sqrt{5})$.

Hence

$P = \begin{bmatrix} -2/\sqrt{5} & 1/\sqrt{5} \\ 1/\sqrt{5} & 2/\sqrt{5} \end{bmatrix}$, $(\phi, \psi) = \mathbf{d}P = (-20\sqrt{5}, 0), p = 5\sqrt{5}$.

Thus, $\lambda_2 = 0$, $\psi = 0$, λ_1 and p have the same sign; this is case 2(c), and the conic is a pair of parallel lines with centre $(4, 0)$.

(e) $A = \begin{bmatrix} 9 & -12 \\ -12 & 16 \end{bmatrix}$, $\mathbf{d} = (20, 15)$, $c = -100$. The eigenvalues of A are $0, 25$, and corresponding unit eigenvectors are $(4/5, 3/5)$, $(-3/5, 4/5)$.

Hence $P = \begin{bmatrix} 4/5 & -3/5 \\ 3/5 & 4/5 \end{bmatrix}$, $(\phi, \psi) = (25, 0)$.

Thus, $\lambda_2 \neq 0$, $\phi \neq 0$; this is case 2(a), and the conic is a parabola, which has no centre.

Exercises 8.5

1 The linear transformation with matrix P is

$T: \mathbb{R}^3 \to \mathbb{R}^3$ where $T((x, y, z)) = (x, y, z)P^T = (x', y', z')$.

Thus $(x, y, z) = (x', y', z')P$

$= (x'/\sqrt{2} + y'/2 + z'/2, -x'/\sqrt{2} + y'/2 + z'/2, y'/\sqrt{2} - z'/\sqrt{2})$.

Substituting for x, y, z in the equation of the quadric Q_1 we obtain the quadric Q_2 which has equation

$$x'^2 + (3 + 2/\sqrt{2})y'^2 + (3 - 2/\sqrt{2})z'^2 = 1.$$

This is an ellipsoid. Since P is orthogonal, T is euclidean, and so Q_1 is an ellipsoid.

2 (a) This is an ellipsoid.
  (b) This is a quadric cone.

INDEX OF NOTATION

$\mathbb{Z}$ integers
$\mathbb{Z}^+$ positive integers
$\mathbb{Q}$ rational numbers
$\mathbb{R}$ real numbers
$\mathbb{C}$ complex numbers

Notation	Page
$\mathbf{v}$	1
$\overrightarrow{AB}$	1
$\mathbb{R}^2, \mathbb{R}^3$	1
$\mathbf{0}$	2
$\|\mathbf{v}\|, \|\overrightarrow{AB}\|$ (in $\mathbb{R}^2, \mathbb{R}^3$)	4
$\mathbf{a} \cdot \mathbf{b}$	9
$\mathbf{a} \times \mathbf{b}$	12
$\widehat{AOB}$	22
$[a_{ij}], [a_{ij}]_{m \times n}$	28
$O, O_{m \times n}$	31
I, I_n	38
A^n ($n \in \mathbb{Z}^+$)	38
A^T	40
A^{-1}	43
A^n ($n \in \mathbb{Z}$)	44
e.r.o., e.c.o.	60
$E_r(\alpha)$	65
$E_{rs}(\alpha)$	65
E_{rs}	66
$\begin{pmatrix} 1 & 2 & \ldots & n \\ a_1 & a_2 & \ldots & a_n \end{pmatrix}$	74
S_n	74
$(a_1 a_2 \ldots a_n)$	75
$P_n(x_1, \ldots, x_n)$	75
$\hat{\sigma}(\sigma \in S_n)$	75
$\det A, \|A\|$	78
$\begin{vmatrix} a_{11} & \ldots & a_{1n} \\ \ldots & \ldots & \ldots \\ a_{n1} & \ldots & a_{nn} \end{vmatrix}$	78
Δ_{rs}	87
A_{rs}	87
$M_{m,n}(\mathbb{R}), M_{m,n}(\mathbb{C})$	104
$\mathbb{R}^n$	104
P_n	104
$\mathscr{S}$	105
$\mathbb{Q}(\sqrt{2})$	106
$\text{Map}(\mathbb{R}, \mathbb{R})$	106
$\langle \mathbf{v}_1, \ldots, \mathbf{v}_n \rangle$	108
$\dim V, \dim_F V$	115
$L(V, W)$	126
T^{-1}	127
$V \simeq W$	128
F^n	129
$S \circ T$	133
$\ker T$	135
$\text{im } T$	135
$\text{rank } T$	138
$\text{nullity } T$	138
$(A\|B)$	147
${}_cA_d$	160
$E(\lambda)$	168
$\|\mathbf{x}\|$ (in $\mathbb{R}^n$)	177

207

GENERAL INDEX

abelian group 3
addition
  of linear transformations 126-7
  of matrices 30-3
  of vectors 2-4
additive inverse
  of a linear transformation 126
  of a matrix 31
  of a vector 3
adjugate 92-4
altitude of a triangle 22
antisymmetric matrix 41-2, 191
associative law 3, 32, 37-8

basis 8-9, 110-16

cancellation law for matrices 39
central conic 193
central quadric 200
centre
  of a conic 193-4
  of a quadric 200
centroid 16-17, 22
change of basis 160-6
characteristic equation 169-71
circumcentre 22
closed 107
coefficients of an equation 52
cofactor 87-94
column-equivalent matrices 60
column rank 142-3
column space 142, 147
column vector 30
commutative law 3, 9, 32, 39
complementary subspaces 116
complex conjugate 185
composite of linear transformations 126

cone 196-7
congruent matrices 180
conic 192-4
consistent system 53-6, 147-51
coordinates 8, 10-14, 132-3
cosine of angle between vectors 177

degenerate conic 194
degenerate quadric 200
determinant 74-96, 169-71
diagonal matrix 45, 166, 184-91
difference
  of matrices 30
  of vectors 4
dimension 8, 112-16
dimension theorem 137-8
distributive laws 4, 10, 12, 32, 37-8
dot product *see* scalar product

echelon form 64-5, 83, 146
eigenspace 168-72
eigenvalue 166-72, 185-200
eigenvector 166-72, 185-200
elementary column operation 60-1, 67-8
elementary matrix 65-70, 85-6
elementary operations on systems of equations 56-9
elementary row operation 60-4, 66-70, 144-6
elements of a matrix 28
ellipse 193-4
ellipsoid 196, 200
elliptic cylinder 198-9
elliptic paraboloid 198, 200
equal matrices 29
equal vectors 1
equivalent systems of equations 56-9

GENERAL INDEX

Euclidean transformation 180–4, 192–201
Euler line 22
even permutation 76
expansion of a determinant 89–91

field 103
finite-dimensional 112
free vector 1–4

Hermite normal form *see* reduced echelon form
homogeneous linear equations *see under* systems
homomorphism 124
hyperbola 193–4
hyperbolic cylinder 198–9
hyperbolic paraboloid 198–200
hyperboloid
 of one sheet 196, 200
 of two sheets 196–7, 200
hyperplane 52

identity matrix 38
identity permutation 75
image 135–50
inconsistent system 53–6, 58
index laws 44
inverse
 of a linear transformation 127–30, 133
 of a matrix 42–5, 68–70, 85–7, 93–4, 133
 of a permutation 75
invertible linear transformation 127–30, 133
invertible matrix 42–5, 68–70, 85–7, 93–4, 133
isomorphism *see* invertible linear transformation

kernel 135–40, 143

leading diagonal *see* main diagonal
left-handed system 10
length of a vector 4, 9–14, 177–80
linear combination 6, 107
linearly dependent 6–9, 109–12
linearly independent 6–9, 109–12, 114
linear mapping *see* linear transformation
linear transformation 124–51, 160–72, 182–4

lower triangular matrix 79

main diagonal 30
matrix 28–45, 57–70, 85–7, 92–6, 130–5, 141–50, 169–72, 177–91
 of a linear transformation 130–5
matrix multiplication 33–40
median ' 16
minor 87

negative *see* additive inverse
nilpotent matrix 45
non-singular matrix 85–7, 149
normal 19–20
normal form for a conic 193
nullity 138–40

odd permutation 76
order of matrix 28
orthocentre 22
orthogonal matrix 177–80, 182–95
orthogonal vectors 10–14, 177
orthonormal basis 184

parabola 193–4
parabolic cylinder 200–1
parallelogram law of vector addition 2
permutation 74–6
perpendicular vectors *see* orthogonal vectors
position vector 4
post-multiplication 66
pre-multiplication 66
product
 of linear transformations *see* composite
 of matrices *see* matrix multiplication
proper zero divisor 39

quadric 195

rank
 of a linear transformation 138–50
 of a matrix 141–50
rectangular coordinate system 10, 192, 195
reduced echelon form 62–5, 68, 143–5
restriction mapping 157
right-handed system 10, 192, 195
row-equivalent matrices 60–5
row rank 142–6
row space 142

209

row vector   30

scalar   102
scalar matrix   45
scalar product   9–12, 177–80
sign of a permutation   74–76
singular matrix   85
size of a matrix   28
skew-symmetric matrix *see* antisymmetric matrix
solution of a system of linear equations   53
span   107–9
spanning sequence   107–9
square matrix   30
standard basis   110, 141
subspace   106–9, 127, 136, 168
sum of *see* addition of
symmetric matrix   41–2, 184–91
systems
  of linear equations   29, 52–9, 94–6, 146–51
  of linear homogeneous equations   29, 94–6, 148

transition matrix   161–6, 184
translation   181–4, 193, 195, 198
transpose of a matrix   40–3, 80, 184–91
triangle law of vector addition   2
trivial solution   95

upper triangular matrix   71, 79
unit vector   10–11, 21, 177

vector   1–22, 102–16
vector product   12–14
vector space   102–16
vector space morphism *see* linear transformation

zero divisor   39
zero mapping   125
zero matrix   31
zero vector   2